P9-CNH-510

ALSO BY FRED ALAN WOLF

Parallel Universes
The Body Quantum
Star*Wave
Taking the Quantum Leap
Space-Time and Beyond: The New Edition
(with Bob Toben)

Fred Alan Wolf

A TOUCHSTONE BOOK
Published by Simon & Schuster
New York London Toronto Sydney Tokyo Singapore

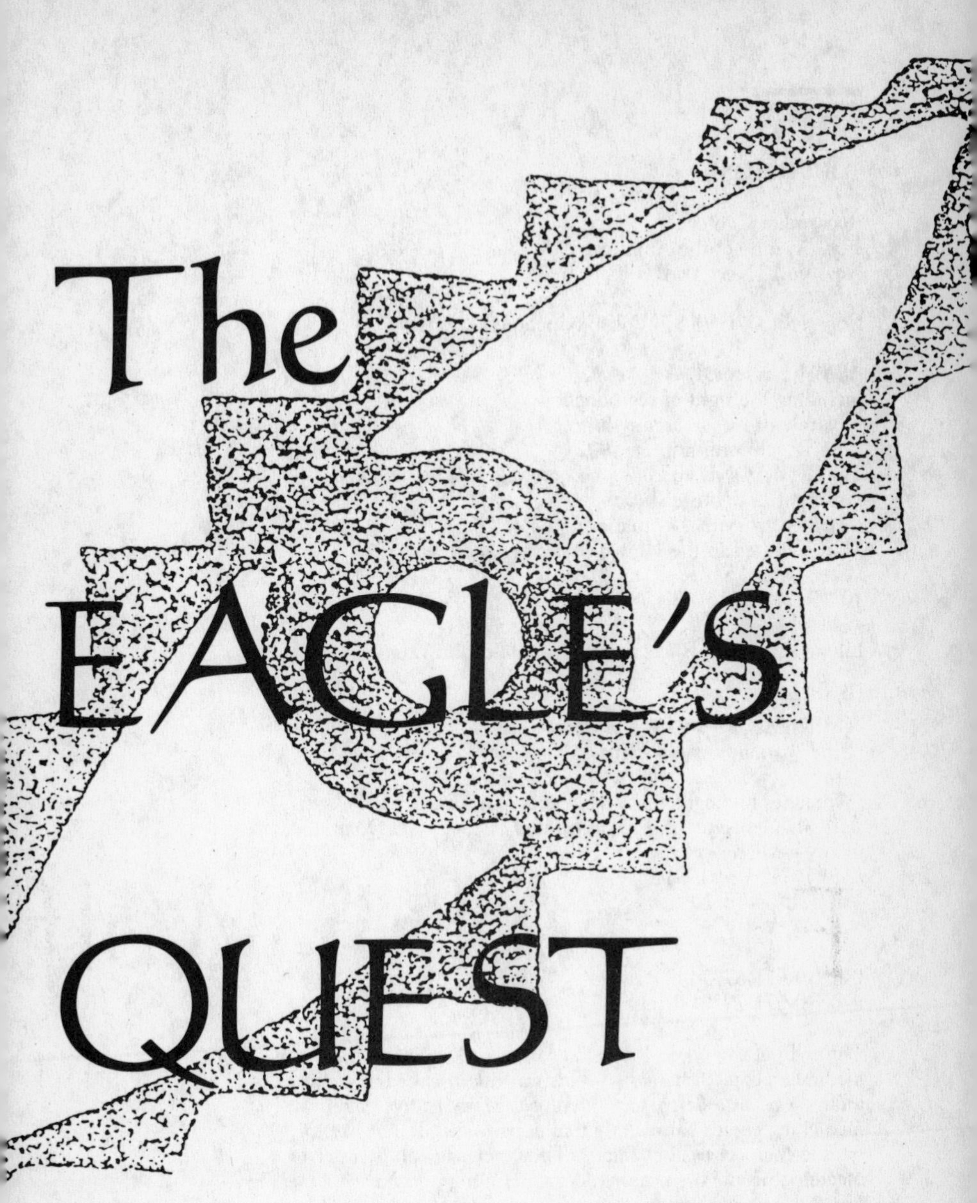

A Physicist's
Search for Truth
in the Heart of the
Shamanic World

TOUCHSTONE

Rockefeller Center
1230 Avenue of the Americas
New York, New York 10020

First Touchstone Edition 1992
TOUCHSTONE and colophon are registered trademarks
of Simon & Schuster Inc.
Designed by Nina D'Amario/Levavi & Levavi
Manufactured in the United States of America

10 9 8 7 6 5

Library of Congress Cataloging-in-Publication Data

Wolf, Fred Alan.
The eagle's quest : a physicist's search for truth in the heart of the shamanic world / Fred Alan Wolf.
p. cm.
Includes bibliographical references and index.
1. Shamanism. 2. Occultism and science. 3. Quantum theory—Miscellanea. I. Title.
BF1611.W84 1991
291.1'75—dc20 91-19046
CIP

ISBN 0-671-67534-6
ISBN 0-671-79291-1 (pbk)

Throughout the book, I have, at times, called Native American medicine people "shamans." This was not intended as a slur or, in any way, as a derogatory inference. Many Native American medicine people, particularly traditional ones, do not like to refer to themselves as shamans. They prefer to call themselves medicine people or, simply, doctors. Among non-Native Americans, the word *shaman* is generally acceptable and used. I have adopted this usage for the sake of simplicity and clarity.

Acknowledgments

In traveling around the world to gather information and meet many people, I found myself owing very much to so many. I have tried to mention the most important friends I met throughout my travels. I may have missed a few names, and to those that I met with, talked to, exchanged ideas with, or even sat next to on a bus from Cuzco to God knows where, and have forgotten your name, please forgive me. To you also I am very grateful for having met you.

I am truly indebted to the following persons. These next words are for them:

To each and every one of you, I owe a debt of gratitude.
You truly are the guts of this book.
Without you there would have been no book at all.
In a very real sense, you, all, are midwives of this book.
It is as much your story as it is mine.

In Switzerland: Holger Kalweit and Amelie Schenk.

In England: Jennifer Maughan and Jerome Whitney from the Druid order; Bridgett Winter; Caroline Wise; Richard Lawrence and Chrissie Aubry from the Aetherius Society; Christopher Hall and Richard Dufton of Brighton; Drs. David Bohm and Basil Hiley of Birkbeck College, London.

In Wales: Paul and Charla Devereux, Dr. Keith and Alena Birkinshaw.

In New Mexico: Maky (Max Ederly) and Edna Oxman, Bruce Lamb,

Jamie Sams and the Heyokah center, Dr. Richard Weise, and Judith Wolf.

In Peru: Dr. Jorge Gonzalez-Ramirez, Raul Espiritos, Don Solon, Dr. Jacques Michel Mabit, Jose Campos, Dr. Fernando Cabieses, Sandra Weise, Nora de Izcue, and Didier Lacaze.

In Arizona: Candace Lienhart.

In California: Kote and Lin A-lul'Koy Lotah, and Terence McKenna.

In Montana: Swain Wolfe and Clara Pincus.

In New York: Bryce Bond and Zulma, and Rhonda Johnson.

In the Pine Ridge Sioux Indian Reservation: Ed McGaa, Doug White, and Paul Little.

In Rapid City, South Dakota: Julie Rencountre and Mick McGaa.

Special thanks are offered to those in the above list who consented to dialogue with me and have segments of that conversation reproduced in part in this book.

I also want to thank my editors, Bob Asahina and Laura Yorke at Summit Books, Simon & Schuster, for many ideas and suggestions regarding the writing of this sometimes tricky interweaving of three storylines into one narrative. Toward the end, Laura was especially patient, wise, and painstaking.

I especially want to thank Candace Lienhart for providing spiritual and psychic protection for me as I made my way through the realms of the shamanic world. I realize now that I needed it. Without it, I may not even be here.

Contents

Cast of Principal Characters

In the United States

Ralph Blum, author in California
Carol Dryer, psychic counselor in California
Judith Gray, my sister in Illinois
Candace Lienhart, American shamanka in Arizona
Paul Little, Oglala Sioux tribal leader in South Dakota
Kote Lotah, Chumash medicine man in California
Lin A-lul'Koy Lotah, Chumash medicine woman in California
Ed McGaa, Oglala Sioux Indian, author and shaman-initiate in New Mexico and South Dakota
Mike (not his real name), translator for Jorge Gonzalez, in New Mexico
Native American healer, name unknown, in California
Andrija Puharich, medical doctor and paranormal researcher in New York
Marvin Red Elk, Oglala Sioux ceremonial singer and drummer in South Dakota
Tony Robbins, firewalk seminar leader in California
Jamie Sams, Seneca Indian shamanka in New Mexico
Scot Sothern, former son-in-law
Snake-woman, dream spirit of sexuality in New Mexico
Doug White, Rosebud Sioux medicine man in South Dakota
Elaine Wolf, former wife, Michael's mother in California
Emma Wolf, my mother (deceased) in Illinois

Jacqueline Wolf, my daughter in California
Judith Wolf, former wife, in New Mexico
Leslie Wolf, my daughter in California
Michael Wolf, my son (deceased) in California
Anthony Wolf, my youngest son in California

In Great Britain

Astral dream woman, identity and location unknown
David Bohm, theoretical physicist in London
Paul and Charla Devereux, spiritual geography researchers in Wales
Richard Dufton, Anglo-Saxon shaman in Brighton
Werner Erhard, founder of est in London
Chris Hall, Anglo-Saxon sorcery researcher in Brighton
John Hasted, physicist in London
Thomas Maughan, former chief Druid (deceased) in London
Nancy, former lover and housemate in London
Jerome Whitney, former pendragon Druid Order in London

In Europe

Jean Durup, physicist in Paris (deceased)
Holger Kalweit, shamanic researcher in Switzerland
Amelie Schenk, shamanic researcher in Switzerland
Carlo Suarés, Qabalist master in Paris (deceased)
Nadine Suarés, wife of Carlo Suarés in Paris

In Peru

Arturo, Peruvian guide in Iquitos
Carlos, graduate student doing plant research in Iquitos
Fernando Cabieses Molina, chief surgeon and shamanic researcher in Lima
Jose Campos, young Peruvian shaman in Tarapoto
Meliton Delgado, a film shaman who in real life is also a shaman in Lima
Raul Espiritos, Peruvian professor in Tarapoto
Franco, Peruvian guide in Iquitos
Jorge Gonzalez-Ramirez, Peruvian shaman in Tarapoto
Ignacio, assistant to Jorge Gonzalez in Tarapoto
Nora de Izcue, film-maker in Lima
Jungle and ayahuasca spirits seen in the Amazon jungle in Iquitos
Jacques Mabit, French medical doctor and shamanic researcher in Tarapoto
Miguel, a film hero in Lima
Nexy, a film heroine in Lima
Nonoy, Peruvian woman and spirit of ayahuasca in Tarapoto

Daniel Pacheco, film producer in Lima
Peruvian passenger on flight, name unknown in Lima
Rude airport spirits of Tarapoto, drunks who were also spirits in Iquitos
Don Solon, maestro shaman from Iquitos
Ted and Lisa (not their real names), Americans in ceremony in Tarapoto
Trisha, graduate student doing plant research in Iquitos
Sandra Weise, film-maker in Lima
Wilfredo, Peruvian shaman in Tarapoto

And, of course, myself, Fred Alan Wolf, physicist and shamanic researcher who interacted with these people in the above countries.

Preface

The story you are about to read is a reconstruction of my interaction with shamanism and my gradual understanding of its overlap with quantum physics, psychology, and modern science in general. To write this story, I spent several years visiting shamans and shamanic researchers in Great Britain, Switzerland, the United States, Brazil, Peru, and Mexico. During these visits, I also went through initiatory and ceremonial experiences. These experiences were led by "new world" shamans consisting of Native Americans (Indians) in North America and especially in the Amazon jungles of Peru. In my earlier experiences with the shamanic world, although it is only with hindsight that I realize that I was in such a world, I spent much time in India, Nepal, Europe, and Great Britain.

The backdrop for my story is based on my experiences in Peru. This is undoubtedly the strangest story of all. It deals not only with my adventure in the mountains of Peru but also with my journeys into the jungles with Peruvian shamans and my actual experiences taking the vision vine, ayahuasca, into my system. Ayahuasca is a substance made from a vine that grows in the high and mid Amazon jungle regions of Peru. It is used in shamanic ceremonies by Peruvian shamans as a purgative. It can also induce visionary states of consciousness.

My experience with the *Ayahuasqueros* (shamans who use ayahuasca) was the most powerful in my investigation. It gave me new hope into unlocking the secrets of the unconscious mind and bringing them to light, heralding for me a new way in which physics and psychology can overlap.

It also gave me new hope for the eventual curing of a number of pandemic illnesses, including drug addiction problems and psychological illnesses in the Western world.

Breaking from the usual type of writing I have done in the past, I have written about my journey into the world of the shaman in a narrative. The narrative consists of three separate stories, all woven into each other. First, there is the story of a movie, which I saw my last day in Lima, dealing with the adventures of two people, who take ayahuasca, that strangely paralleled my own story. Second, there is my own adventure story consisting of where I went, who I saw, and what I did. And third, there is the developing idea of a new physics of consciousness and how it explains some, but not all, of the extraordinary events that happened to me.

Although together the three stories could appear to be pure fiction, they constitute the truth as I saw it. Indeed, even as I began my adventure, I realized that I was actually on a quest for truth, although at times I wondered if I was even able to separate truth from fantasy. In this quest for truth, I hoped to use my understanding of modern physics, psychology, and the human body to explain shamanic powers. My background, both in quantum physics and in performance magic,[1] I believe qualified me for my task.

I suspected before I wrote this book that some of the principles of modern science, particularly the observer effect[2] in quantum physics, could indeed apply in the healing art world of the shaman. Perhaps shamans have been using in their practices, although conceptualized in ancient world-views and systems, such principles.

Well, sometimes I was successful in proving this theory; other times I was not. Shamans were using principles of modern science and they were using other things, even more real and bizarre than quantum physics. These things I could not reconcile with ordinary causality as described in physics. Surprisingly, I could reconcile them with a new vision of determinism or, if you would rather use another word, fate, and something else even more mysterious and yet so simple that, I believe, everyone will understand. I ultimately learned things about primal and shamanic vision that I simply was unaware *I already knew*.

[1]I not only have a Ph.D. in theoretical physics, but have also worked professionally as a close-up and stage magician. I still do tricks from time to time even though I no longer perform professionally.

[2]The observer effect is the sudden change in a physical property of matter—particularly at the atomic and subatomic level—when that property is observed. This is measured by the change in the probability of observing that property. When it occurs, something that was possible suddenly jumps and becomes actual. It is ascribed to the actions of consciousness upon material objects.

I want to emphasize that what I was seeking was truth, and nothing but. I didn't want my account of events to appear as if I were constructing the truth to please the reader. But even though I have endeavored to tell the truth about my adventures in this strange world of shamanism to the best of my ability, I need to explain that I was a participant and not an objective observer. In a true quantum physics perspective, the observer and the observed became one. I was the observer and at the same time I was observing myself. This was difficult to anticipate or intellectualize about. The experience of it was something I had never had before and was not as I had preconceived it.

Indeed my physics training and its emphasis on materiality and objectivity sometimes got in my way to the truth. Let me explain. Much of today's science is based on a materialist philosophy that is also the basis of Western medicine. People get well because they take antibiotics. Cause leads to effect, and all is well. Shamans may also give "medicine" to the sick, but often the medicine is a placebo. The real healing takes place in a manner not explainable by materialist science.

WHAT DOES THE WORD SHAMAN MEAN?

Most of us in the Western world have been exposed to shamans, but not in the way that really shows what they are. If you just think of a movie taking place in an African or South American setting, you might remember a scene where a witch doctor casts a spell on some poor wayfaring wanderer who happened to come into the tribal village. That witch doctor or voodoo spell-caster was Hollywood's way of presenting, to the modern world, a view of a shaman.

The original word *shaman* came from the Ural Mountains in Russia. It applied to people who acted in several nonordinary capacities for their tribes. It was these capacities that interested me and mark the following chapters of this book.

All shamans have the ability to use sound and vibration in apparently magical ways to alter consciousness (Chapter 3). All shamans go through an initiation period before they become full-fledged shamans. Usually the shamanic tradition has been passed on from elder to initiate. All shamans have teachers who were shamans (Chapter 4). They all know the power of sacred places and plants. I later discovered that there is a connection between sacred places, plants, and where shamans appear in the world (Chapter 5). They also know how to use sexuality to enhance healing and potency in people (Chapter 6). They all have healing power that is based on some form of vibrational energy (Chapter 7). They are all able to enter into a trance state and, somehow, visit other, possibly parallel, worlds. They can, in a way that seemed inexplicable to me when I began my journey, change their bodily forms when they enter parallel worlds (Chapter 8). They also see into time, either the past or the future, of individuals or of the tribe. Often, in the past, they would be used in helping the tribe find new food sources or new places where they could live, if they were a migratory tribe. They are generally visionary (Chapter 9). I also found out that shamans were probably born with these outstanding abilities. They also had to go through a near-death experience in order to develop them. The death realm was something that they dealt with in everyday terms. They were able to access that realm at will (Chapter 10). Finally shamans are able to alter their consciousness to see into other worlds and to heal people (Chapter 11).

In each chapter, I have woven together what I learned from the Old World shamans of Europe, what I both learned and experienced with the New World shamans of North and South America, and some form of explanation based on my understanding of quantum physics, consciousness, and psychology.

Because of my training in physics, and the reputation that physicists deal only with reality, you might be curious about the imaginative writing I sometimes resort to. I have included descriptions of visions I had with Ayahuasqueros and when participating in ceremonies with Native American shamans. I also have included descriptions of relevant dreams.

The episode about the hunting eagle in Chapter 4 and the Inca ceremonial sacrifice fantasy in Chapter 9 came to me in a vision when I first took ayahuasca with a Peruvian shaman. Months later I saw a similar ceremony occur with the sacrifice of a llama when I observed the Inti Rahmi in Cuzco. The episode describing an initiation ceremony in Chapter 4 where I felt the eagle's presence was based on an actual experience I had with an Oglala Sioux Indian shaman in New Mexico. The sexual dream sequence in Chapter 6 was based on an actual dream I had a few weeks before I met the Peruvian shamans and, indeed, before I even knew I was going to spend so much time with them. The connection to them was more than coincidental. The lucid dream sequence in Chapter 8, where I was taken to an astral level of souls that had committed suicide, took place many years ago, well before I knew I would be writing this book.

I felt that including these fantasy episodes and dream material in this manner was instrumental to understanding my quest and discoveries. It enabled me to capture the feelings I had during this investigation as well as the logical progression of my thoughts.

This was a journey into myself in a way I was not expecting. I believe that I have discovered part of the secret of shamanic healing. What remains to be discovered can only be learned by my further interaction with the shamanic world. I am both looking forward to this and fearing it at the same time. Although this story is finished, I still don't feel that my shamanic journey is over. The unknown still waits for me and beckons me to its darkest corners.

And now I invite you to take the journey with me. I am going to take you back in time, step by step, through each ceremony and through each experience that I had. I want you to see just what I had discovered, when I discovered it, what I was able to piece together in terms of a consistent picture related to quantum physics, and what I felt was missing all along my journey. It was this final piece of the puzzle that I grasped the very hour I ended my journey.

I

An Intuitive Physicist's Quest

It was three in the morning and I was in a hostel room in Lima, Peru. I had awakened early and my mind felt as if it were on fire. I was sitting up in bed waiting for the dawn and a taxi to take me to the airport back to the United States. Although my story begins here, my shamanic journey had seemingly come to an end.

It was the end of June in 1989 and I had been here in Peru for nearly a month. I had hoped to find out how shamans, usually found in what we Westerners call "Third World countries," heal. Although I had gathered a lot of data and had gone through a number of mind-boggling experiences, I still wasn't sure.

I had always wanted to know the secrets of the universe. I had been trained as a physicist and learned to respect my profession. It had taught me that there were no free lunches and everything in the world was accountable. The laws of physics were not to be violated. Energy had to be conserved, and reality was something tangible, even if we could not know its ultimate structure. Seemingly, magic was gone, displaced by the carefully controlled scientist's experiment that was always repeatable. There were no errors in the scientist's world, only badly performed experiments. The only way forward, I thought, to understanding the universe was through the doors that modern physics opened.

Sitting on the edge of my bed, gradually awakening to my rational self, I recalled the thoughts I had when I began my quest. I knew that even though I had studied quantum physics and knew of its strange predictions, I was still very much a Newtonian thinker. I had a machine inside of me running as logically as the gigantic universal clockwork of Newton's imagination. It was my own mind.

Things that constituted the shamanic world such as magic, spontaneous healing, people changing into animals, seeing into the future, and other paranormal phenomena were, to say the least, impossible. Yet, I had heard stories from many people that there were shamans on the planet capable of creating real magic. So I decided to find out if there really were any such beings alive and working today. And if there were, I wanted to learn their secrets. As it turned out, I uncovered more than even I had bargained for.

I looked out of the window. It was still dark and I was literally searching through the darkness of my own soul. Thinking over all of my experiences, I realized what I had learned and what little knowledge I had when I first began this project. I thought that I had understood just what it was that I would be investigating. I had a theme in mind: the relationship between the practices of shamans and the principles of modern physics. I saw myself as an objective investigative "reporter," like Joe Friday in the television series *Dragnet*, just "seeking the facts, ma'am."

I didn't really expect to get involved with the shamanic world as much as I did. I found, however, that I really had no choice. I discovered, as I probed, that there was no way for me to determine the relationship between the shamanic world and the world of modern science without involvement. Indeed it seemed that I had to go through a number of processes that required my participation with shamans. Because of that, I can't really say that I was just an objective observer, "recording the facts, ma'am." I became a participant and that sometimes gave me great difficulty, including physical discomfort, illness, and even fear that I would not recover from the venture. I now see that I had journeyed to another reality, a primal world, another way of "seeing." I became a shaman-initiate without really knowing it.

SHAMANIC PHYSICS: NINE HYPOTHESES

Gradually, with each step I took as I journeyed into the shaman's world, I began to form a somewhat loose set of hypotheses of the relationship between modern physics and shamanic practices. It is really only with hindsight that I can recall my thoughts.

The first hypothesis was: *All shamans see the universe as made from vibrations*. I had learned about vibrations throughout my physics education.

Vibrations are repetitive patterns that can be seen in the simplest physical systems. From the movement of sound through the air, to the motion of a ship rocking from starboard to port, to the invisible vibrations of light waves speeding through the universe, all are examples of vibrational motion.

In my study of quantum physics I had discovered that there were even more subtle vibrations present. These were contained in the "probability" waves of subatomic and atomic matter. These waves are called probability waves, *quantum waves*, or *quantum wave functions* because they have a vibrational pattern and they determine just how probable physical events are to take place. Wherever in space and whenever in time an event manifests are governed by the strength or amplitude of these waves.

From these "vibrations of probability," physicists are able to determine the sometimes mysterious behavior of matter and energy. According to this "new physics," everything in the universe has this inherent probability-vibrational pattern. This pattern enables physicists to calculate the very structure of atoms and molecules and how these particles emit and absorb electromagnetic energy.

Because these waves are nonmaterial, they have bizarre properties. When two particles of matter interact, their probability waves entangle so that no clear separation between them takes place, even though the particles could be miles from each other after the interaction. In this sense although the particles are separated in space, they are still connected. In fact any observation performed on one of the particles instantly, faster than light, produces an effect on the other.

Shamans, although few of them understood the principles of quantum physics, believed in a similar vibrational understructure of the universe. Thus quantum physics, like the shamanic belief, indicates that the universe is also made from vibrations and that everything in it is connected by these vibrations.

SHAMANIC PHYSICS AND MYTHIC REALITY

Quantum waves are invisible. They are constructs of human thought necessary for our modern world to enable us to understand atomic and subatomic matter. Nevertheless, even though we believe in them, we have never actually observed them. They are part of a physicist's mythic system. Mythic though they be, they are vital. Without them there is no way to understand our universe. In a very real sense, quantum waves are ghosts in the machine of reality. They exist for physicists as spirits exist for shamans.

Yet, I realized that today few people living in the modern world believe in spirits or shamanhood. But shamans did. They saw the world differently than I did. The second hypothesis concerned the fact that somehow *shamans were able to see the world in terms of myths and visions that at first seemed contrary to the laws of physics*. They were able to see beyond the usual barriers that inhibit our Western minds. What did shamanic visions consist of? How were they created? And did those shamanic visions form the basis for our stories and myths?

I suspected that the mythic level of reality, which has been written about for millennia, could be found in the shaman's perceptions of past and future. Perhaps shamans see mythic pictures because those pictures are superpositions of events taken from a culture's pasts and futures.

How can anyone see into the future, you might ask? I had some ideas about that. I had written about them in my previous book *Parallel Universes*. In brief, according to the transactional interpretation of quantum physics,[1] these invisible quantum waves of probability originate in the present, in the past, and in the future. For any event to manifest, these waves coming from the future and the present or from the past and the present must interfere with each other in the present. The pattern of that interference then creates matter and energy as we perceive them. Somehow shamans were able to see to either the past or the future sources of those waves. In this manner they were able to construct visions that had mythic proportions and appeared to them as archetypes in the Jungian sense.

ALTERED CONSCIOUSNESS

All of us dream. According to Jung, many times our dreams are filled with archetypal images. For example, we all at one time or another dreamed of flying through the air unaided by any mechanical device such as an airplane. I suspected that shamans were able to alter their waking consciousness invoking archetypal images at will.

I suspected that all of us were capable of doing this, but to do so we needed, as a culture, to learn to alter our consciousness—a very touchy matter for us. Certain cultures throughout history have used mind-altering substances for spiritual and visual guidance. Our culture seems to have lost the purpose but hung onto the substance, manifested in our desire for drugs

[1]See the references to the transactional interpretation under John G. Cramer in the bibliography.

and alcohol. This sobering thought made me recall the third hypothesis: *Shamans perceive reality in a state of altered consciousness.* I suspected that a role for ordinary consciousness was required in any correct model of quantum physics. One could not escape the observer effect—wherein the choice of an observer to measure a particular property of a system forces the system to emerge from a probable state into an actual one. The observer of a quantum system disturbs the system by observing it. Was the act of observation a quantum event? If it was, then consciousness was capable of being understood in possibly the same way we understand matter, through quantum physics. Since shamans deal in vastly different states of consciousness, perhaps there was a way I could understand these states in terms of quantum physics. Perhaps shamans manipulated matter and energy by some form of observational power brought forward when they were in an altered state of awareness.

THE SHAM OF THE SHAMAN

Following along the path provided by the observer effect, there was a fourth hypothesis: *Shamans use any device to alter a patient's belief about reality.* The old adage says "Seeing is believing." Shamanic reality turns it around. People only see what they believe. They will work within the belief structure of a patient in order to pull him or her out of it. In healing a stubborn patient, they will create a trick to alter the patient's fixations. I mention this because, as I discovered, shamans are also tricksters. They will perform conjuring tricks to convince the uninitiated that the trickster has such powers. I had sensed that shamans used some form of trickery, but that trickery may not be what it seems to be to a skeptic. Indeed, it might just be that "trickery" is the observer effect in action. Both the shaman and the healee must be convinced that shamanic power exists even if "trickery" is used on a particularly stubborn patient.

THE PHYSICS OF MEANINGFULNESS

This led to the fifth hypothesis: *Shamans choose what is physically meaningful and see all events as universally connected.* The clue seemed to lie in how they chose the reality they perceived to be true. The answer was they chose to pay attention to those events in their lives that they felt were meaningful. But what did that mean? What is meaningful? Is there a quantum physics of meaningfulness?

I believed that meaning arises as the relationship between events and that somehow shamans could sense a relationship between *any* two events as if they were the ends of a strand of a spider's web. The Anglo-Saxon shamans saw the universe as a gigantic vibrating web. Some physicists, such as David Bohm, saw the universe as a hologram.[2] Perhaps the hologram and the spider web were the same thing seen from different cultural viewpoints.

A web is made of vibrating threads. A hologram is made from light waves, vibrations of energy. By capturing waves of laser light which were emitted from a scene containing objects and causing them to interfere with waves from a pure reference wave of light, information is caught and frozen in a film emulsion. When another reference wave is shone again through the film the original scene is viewed in full three dimensions.

Bohm calls the hologram the *implicate order*. This order is normally invisible but yet contains all of the possible phenomena that can be experienced. When an experience occurs, the order is changed. This new order he calls the *explicate order*. Thus what is explicate is what is observed.

PARALLEL SHAMANIC WORLDS ARE REAL

The sixth hypothesis was: *Shamans enter into parallel worlds*. According to one interpretation of quantum physics called the "many worlds interpretation," which I also wrote about in my book *Parallel Universes*, there are other worlds present that affect our world. Now by a "world" I meant a set of experiences that take place within a region of space and over a period of time. That spatial region could be very large or quite tiny—as big as a room or as small as a muscle fiber. Thus, these events could be taking place in the brain and nervous system. Similarly, the time interval in which these experiences occur could be as long as days or as short as a few thousandths of a second—the time it takes to send a neural impulse down the axon line.

Accordingly, this world we commonly experience is actually a multiple reality—a composite of many other realities. Some of them are very probable—and thus if we have access to them we notice no differences from one to the other. These nondiffering realities form what we call our common past experiences, our memories. Such a multiple reality composed of nondiffering simpler realities we call, simply, normal reality—the world as we see it.

But some of these other realities are not so probable. We thus usually fail to pay any attention to them. They are off the beaten path, so to speak.

[2]Bohm's views are expressed in several books. The best is: Bohm, David. *Wholeness and the Implicate Order*. Boston: Routledge & Kegan Paul, 1980.

When we experience them we say that we have had access to an extraordinary reality or set of experiences.

These extraordinary realities involve processes that are not very likely, and if they were to occur in our common reality we would say that the laws of physics, at least those laws as described by Newton and Maxwell—the so-called classical laws of physics—would be violated. But in quantum physics we must take into account these bizarre realities in order to successfully explain the simplest atomic and molecular processes.

Shamans were aware of a number of different, improbable parallel realities. These included out-of-body experiences, shape-shifting and changing into animals, and time traveling both to the past and to the future. I believed that their nonordinary realities or parallel worlds were relatable to what I had discovered in physics.

A SENSE OF A HIGHER POWER

My mind was racing. I got up from the bed and looked out of the window again. The sky was beginning to lighten. It was another gray day in Lima. Soon I had to leave for the airport. Okay. I had somewhat of a quantum physical basis for how a shaman heals. It consisted of quantum vibrations, the observer effect, and parallel worlds. Together this equated to shamanic healing. Could I construct a model? This model would explain how the shaman and the patient reached an alternate state of consciousness together. To make a model I needed the seventh hypothesis: *All shamans work with a sense of a higher power*.

What was that higher power and how did it manifest? I knew that shamans see themselves as part of the great universe-mother. They felt that they were able to tune to a higher power, the mind of God, perhaps. They were able to reconstruct reality by using that power. But where did this higher power come from? Of course, we all know about the higher power associated with heaven and the lower power associated with hell. But I suspected that shamans dealt with powers that lay between these two extremes.

Here I had an idea: the higher power they felt came from somewhere in between—the earth itself. Shamans were, in a sense, materialistic. They used the planet to enhance their healing and magical powers. Perhaps *used* is too strong a word. They connected to our planet by using sacred plants and living near sacred places. In fact I suspected that shamans got their powers from sacred places and plants.

SEXUALITY, LOVE, AND HEALING

And there was another power I had sensed with all of the shamans I had met. This power was not necessarily the same as the higher power mentioned above, although perhaps it is. All the shamans I had met with appeared to me to possess extraordinary abilities to love and be compassionate. I suspected that this was no accident. It was necessary for them to use the power of love in order to heal. This led to the eighth hypothesis: *Shamans use love and sexual energy as healing energy*. I knew that shamans practiced male and female medicine. They saw that every human being was psychically both female and male and that an imbalance in sexual energy often resulted in illness. How was I to connect this with physics?

I knew that quantum waves existed spread out through space and even beyond the barriers of time. I knew that these waves affected matter. In the Bohr interpretation of quantum physics these waves would vanish the instant any observation of matter occurred. In fact, the observation was pictured as the sudden collapse of the wave producing a particle of matter. I then pictured the wave as a feminine action and the particle as masculine action. I had first written about this in my earlier book *Star*Wave*.

I also suspected that sexual energy had much to do with the interaction of light with matter. According to the Pauli exclusion principle, electrons have the ability to exclude each other from entering each's territory. Atomic energy structures are possible only because of electronic exclusion.

Also according to Wolfgang Pauli, the Nobel Prize-winning physicist, photons, particles of light, have the ability to "include" each other. They are able to enter each other's territory and in fact have a strong tendency to do so. Lasers operate because of this photon inclusion.

Thus love was photon inclusion, while isolation and separation arose through electron exclusion, and sexual energy was the dance of light with matter involving the music of photon inclusion and electron exclusion.

Perhaps, I thought, shamans were tuning to this magical aspect of quantum physics: they knew how to touch on photon inclusion and electron exclusion in order to heal and to bring forth their own magic and powers.

THE TEST OF DEATH

The last and ninth hypothesis concerned how shamans changed or altered their consciousness. It frightened me to even think about it. *Shamans enter the death world to alter their perceptions in this world*. My experiences testing this hypothesis were telling me that this was true to the best of my perception.

But why was death so necessary for life and healing? It seemed paradoxical to me. Yet I had the uncanny feeling that this was probably the most important discovery I had made. But what connection did this have with quantum physics? I began to suspect that it had something to do with the observer effect. By altering the way I was observing myself and my environment, I was changing my perception. By becoming aware of processes that were not chiefly concerned with my bodily survival, I was able to see other realities.

A RECAP: SHAMANIC PHYSICS HYPOTHESES

The key insight in all of this connecting quantum physics to shamanic reality and connecting each part of my hypotheses was found in the observer effect of quantum physics. Alter the way you see reality and you alter reality. Both quantum physicists and shamans did this, although in entirely different ways and with entirely different viewpoints. Because it will be helpful for the reader in going through my journeys in the following chapters, I will recap the nine hypotheses:

Hypothesis 1. All shamans see the universe as made from vibrations.
Hypothesis 2. Shamans see the world in terms of myths and visions that at first seemed contrary to the laws of physics.
Hypothesis 3. Shamans perceive reality in a state of altered consciousness.
Hypothesis 4. Shamans use any device to alter a patient's belief about reality.
Hypothesis 5. Shamans choose what is physically meaningful and see all events as universally connected.
Hypothesis 6. Shamans enter into parallel worlds.
Hypothesis 7. All shamans work with a sense of a higher power.
Hypothesis 8. Shamans use love and sexual energy as healing energy.
Hypothesis 9. Shamans enter the death world to alter their perceptions in this world.

As you read my story I will, at the beginning of each chapter, bring forth these hypotheses as they apply to my discovery. Sometimes more than one hypothesis applied to my experience, and sometimes I was not successful in relating physics to the experience I had with these remarkable people.

I made my way downstairs and walked out of the hostel. I was on my way home. When I entered the cab, I suddenly saw the whole story of shamanism flash before my mind. I was beginning to see more clearly how the story connected together with my quantum physics understanding.

But yet, even though I had entered shamanic reality and had made connections, I was still missing something—something very obvious and yet elusive at the same time. What was it? I felt it as you might feel when you have a word on the tip of your tongue and can't quite get at it. Or when you sense something in an uncanny way that you know that you cannot know. My mind returned to the previous day, just fifteen hours ago. It seemed more like a lifetime. As the cab made its way to the airport, a strong wind came up from nowhere and that uncanny feeling arose again. I smiled and my mind drifted back to the previous day. . . .

2

The Winds of Ayahuasca

It was three in the afternoon. I was upstairs in my hostel room waiting for a Peruvian friend to come and fetch me. I realized that after all of my experiences in the shamanic world, I was still confused. I still didn't know just how to put my experiences together in a cogent form. I still didn't know just how quantum physics fitted into the story.

I was feeling a little frustrated, but in a few minutes my friend Sandra Weise, a Peruvian film-maker, was coming to take me to a movie. I didn't know then that the movie would actually solve my problems and bring my quest to an end.

At 3:10 P.M., Sandra showed up. Sandra was a good friend of the film's director, Nora de Izcue, whom I had met earlier. In the car with her was Daniel Pacheco, the producer of the film. We made our way from the suburb of San Ysidro to Lima Centro, the downtown of the city. After our arrival, we were directed to a small viewing room in the basement of a large theater. The three of us sat down. I asked Sandra if she had previously seen the film. She hadn't.

The movie was shot in 35 mm and in color, to run for about 90 minutes. It was entitled *The Winds of Ayahuasca*. It told a strange love story that revolves around the magic-religious beliefs of the people in the Peruvian Amazon. The hero, Miguel, a young and handsome professor of sociology in Lima, well into the Western way of life, is on vacation in Iquitos, a jungle city at the mouth of the Amazon River.

Miguel finds the waterfront of Iquitos fascinating. He realizes that although he is a Peruvian, the Indians and mestizos of Iquitos call him *gringo*. As far as they are concerned, he is from another world.

As he wanders the waterfront, he meets by chance a strange and lovely woman named Nexy, who was living in a slum and worked occasionally as a prostitute.

Nexy is haunted by spirits called *yacurunas*, which are phantoms arising from the Amazon River and are believed to be the pink dolphins found there. Miguel, strangely attracted to the woman, decides to help her. They go to see a healer (played by Meliton Delgado, the son of the famous shaman, Jose Delgado). Miguel, although reluctant to get fully involved, watches the ceremony as Nexy takes the ayahuasca—the vision vine—to purge herself of the spirits. The film shows the ceremony and, based on my own experiences with this substance, appeared quite accurate in what it depicted.

Nexy, having taken the ayahuasca, and Miguel follow Meliton, the shaman, to a jungle retreat, and there she relives her past and her relationship with her mother, who died when Nexy was very young. She sees how the death of her mother and her early childhood contributed to her spirit haunting, but she is still incapable of ridding herself of the spirits.

The shaman realizes that Nexy needs more help, and he leads the two characters deeper into the jungle (a metaphor for penetration into the unconscious) in order to obtain the ultimate cure for Nexy. The shaman encounters the yacurunas and manages to drive them off, for a little while. But as he goes deeper into Nexy's problems, the black magic of the spirits gets stronger. A clash occurs between the black and the white magic, and Nexy disappears without a trace.

Miguel then takes ayahuasca and begins to search for Nexy. His search brings him closer to the shamanic world that Nexy inhabits, and he learns in this dramatic way that such a strange world has now become part of his own world.

As I watched the film I became visibly moved, nearly on the brink of tears. The story of the bearded gringo academic going into the jungle and finding a woman who takes ayahuasca struck me as being parallel to my own experience. I, too, flew into the jungle, to the Amazon River town of Iquitos, where I found a beautiful, mysterious woman who was to be my guide into the mysteries of ayahuasca and jungle spirits. Her name was Nonoy. The fact that in the film Miguel takes ayahuasca and it alters his world view also paralleled my own.

I was also struck by another aspect of the story. I had come to Peru to visit Jorge Gonzalez, a Peruvian shaman I had met earlier in the United States. In the film, Jorge, the shaman I had come to see in the jungle, surprisingly appeared as an actor. He played the role of a playboy who invited Miguel to Iquitos. I was quite surprised to see Jorge in the movie and was impressed that he was such a good actor. Jorge, in my story, in so-called real life, invited me to the jungle. The appearance of Jorge as an actor in the film resonated with my actual experience. Jorge, my shaman and my guide into the ayahuasca ritual and the main reason I even came to the Peruvian jungle, was also the main actor in my "film"—my voyage into the jungle and to the city of Iquitos.

But this was only a movie, my rational time-bound mind said. They were only actors playing out roles written down by a scriptwriter. Yet, I had the strangest feeling that I, too, was on a journey that was somehow scripted, written down, and I was just following that script, that storyline, so to speak. Nora's film was a final reminder, a fitting capstone, that my story was, too, just a play and I was only an actor. Only this play was written for me (by whom?) so that I could find out things about myself that I didn't want to look at before.

As I watched the film, I thought back to every scene in the life of my story, from the time I began this book project, from the time I entered Peru, until the day I was to leave it. My life seemed more and more like a movie script, a play, a story in which I was guided by the wind of some moving finger of the writer (whoever she or he was). I realized that I was not able to sustain my own force of being or identity. Everything that happened to me was just a movement through a script. There were too many synchronicities, too many events that I could not control or alter.

But my life movie was not contained on a two-dimensional flat strip of celluloid cut into a sequence of frames running through a projector. It was my life in full living three spatial dimensions and at least one time dimension. It was more than a film. It was closer to a hologram and I was experiencing it.

As I searched back through my memory, many, many recollections flooded my mind. Suddenly scenes from my childhood flashed before me. I realized that my journey into the heart of the shamanic world began in my childhood with my interest in magic. I remembered having a great interest in magic—the kind that Harry Blackstone used to perform at the Chicago Theater just before the film started at that downtown movie house many, many years ago.

I remembered a particular day when I was playing in the front hallway of my apartment building.

I stood at the top of the stairwell and looked down wondering if I could fly down the nine or ten stairs reaching to the ground floor from our first-floor apartment. Without thinking I remember skidding down the stairwell with my feet only barely touching the leading edges of each step. I was on the ground floor in a flash, and I hadn't slid down the bannister nor had I placed my feet on any of the steps. I was barely eight years old.

When I grew older and thought of what I did, I realized it was impossible. My feet just weren't long enough to go from one step edge to the next without my falling flat on my face. Was this just a dream of super powers or had I actually skidded down those stairs? To this day I don't really know.

Through my early years I maintained my interest in magic and fantasy. That interest carried me into thinking about the world a little differently from my fellows. It led me into physics.

By 1957, I had graduated with a BS degree in engineering physics, and those were the days of atoms and roses before the feet of scientists. The atom bomb had proved to the world that the United States was a power to respect, and everyone believed that science was the answer to the world's problems.

Soon I enrolled in graduate school, and by June of 1962 I had completed all the requirements for the Ph.D. in theoretical physics. By February of 1963 I had become Dr. Wolf.

But I paid a price for my learning. A dear price that I hadn't even realized at that time. Somewhere in all of that education, I had lost the magic. I simply accepted the physics education as an indoctrination. I can't say I really understood physics even after getting my Ph.D. I knew how to manipulate the math and logically prove whatever needed proving according to the rules of physics, but I hadn't understood it at all.

I had become a kind of physics machine, and even though I did have some creative moments solving academic problems, I really didn't know how physics fit into the total scheme of things.

After I received my degree, I took an academic post at San Diego State University. Although I taught there for several years I began to feel a growing uneasiness inside of me. I felt that something was missing in my life.

By the late fall of 1970, I found myself, as part of a round-the-world sabbatical leave, traveling through India. Life in the Indian subcontinent impressed me in subtle and strong ways. I was also beginning to realize that the world could not be figured out rationally—something that my physics background would never have allowed me to believe. By spending time in India, I was literally discovering another world, another way of thinking, believing, and living. I wasn't equipped to handle what I was experiencing. It seemed more like magic to me. But this wasn't the last bit of magic to enter my life on the Indian subcontinent.

In October, I flew to the Himalayas and visited the city of Katmandu in

Nepal. One bright but cold sunny day, I left my hotel room and wandered through the village not really knowing where I was headed. Soon I had left the center of the village where I was staying and climbed a hill. Peering upward at the top I had noticed a prominent obelisk with a pyramidal peak. On each face of the pyramid was a single painted eye—the eye of the universe peering out into all four directions. With that goal in sight, I climbed and was surprised when I saw that a lovely Buddhist temple was on the same grounds as the obelisk.

When I reached the top it was past midday, perhaps around four in the afternoon. The sun was beginning to sink low in the sky, although there was still plenty of light. The air was strangely silent and still. I could see many monkeys playing around the rooftops of the temple, but they too were quiet. (Later I visited the same site to find the monkeys chattering quite loudly.) Finding that the door was open, I ambled into the temple and stood for a while in the vestibule. Peering inside, I saw a long line of vacant cushions seemingly waiting for people to sit on them. I was unable to see the wall at the end of the room; the line of cushions seemed to stretch out to infinity, and for a brief moment I felt as if the place were totally timeless.

Then suddenly there was a low brassy sound, like someone playing a bass note on an elongated French horn. Just as quickly people began to appear. They all had bald heads and wore traditional orange garbs. I soon realized that these were Buddhist monks taking their seats on their cushions to begin their meditation.

After they had taken their places they became silent for a brief time before they began to chant. The chanting was totally hypnotic. I couldn't find another word to describe the effect it had on me. While standing and listening intently, a fly landed on my foot. And just then—I can't explain it any differently—I became the fly. I actually felt myself at one with the fly. I couldn't sense any separation from it. I could still see it on my shoe, but, seeing my shoe at the other end of my leg, it seemed miles away.

Even though my sight was telling me that everything was just as I rationally expected it to be, all of my other senses were telling me something was quite different. How could I be a fly? Yet that is what I felt. (Writing about this now, it seems stupid. My first transcendental experience and it's with a fly!?) Meanwhile the chanting continued and I became rapturous. I was swooning in a sea of Buddhist waves of sound and I was one with the fly.

SCIENCE HINTS

How could I have felt this? In my study of quantum physics, I later found a hint that this was explainable—in an experiment that was performed in 1982. In the experiment, described theoretically by Albert Einstein, Boris Podolsky, and Nathan Rosen as early as 1935, and later actually carried out in slight modification, by a French physicist, Alain Aspect and his group at the University of Paris in Orsay[1] in 1982, two particles were separated by more than ten meters. Yet independent measurements performed on them indicated that some form of connection had to exist between them even though there was no material force connecting them.

Even though it is possible to write down the mathematical expression explaining this connection, it doesn't give you the feeling that it is true. Even though those simple particles are separated from each other, they remained in a single state of togetherness. Even though I and the fly were separated, we were also connected.

The experience was a hint, a harbinger of something I was to feel again and again as I made my way into the shamanic world. There was some form of connection that existed between us all. At this point, I didn't know what that connection was. I had only the theory from quantum physics to back up my feeling. I didn't know it then, but this experience was shamanic and fit with my hypothesis that shamans see all events as universally connected.

Then it dawned on me that I had felt this connection when the horn sounded and the monks began their chants. Somehow the sounds in the Buddhist temple shifted my awareness away from myself. Somehow I had become aware of another way of seeing and being in the universe. I had discovered that there was a magical power in sounds and words.

[1]See Einstein, Albert; Podolsky, Boris; and Rosen, Nathan. "Can the Quantum-Mechanical Description of Physical Reality Be Considered Complete?" *Physical Review*. Vol. 47, p. 777, 1935, and see Aspect, Alain; Dalibard, Jean; and Roger, Gerard. "Experimental Test of Bell's Inequalities Using Time-Varying Analyzers." *Physical Review Letters*. Vol. 49, No. 25, p. 1804, 20 December, 1982.

3

The Power of Words and Sounds

In the theater with Sandra and Daniel, watching the movie and at the same time remembering the movie of my own mind, I had an overwhelming sense of destiny that my quantum physics training could not explain. It was as if I were just a leaf blown by the winds of ayahuasca—the title of the film that I was watching. Time seemed to make no sense. The Newtonian clockwork of my own mind had come to rest. It was silent.

In the movie, the shaman conducting the ceremony was singing strange and enchanting songs to Nexy and Miguel. These chants, just as the chants I heard when I took part in my own shamanic ritual with the Peruvian shamans a few weeks earlier, brought all of my own experiences with words and sacred sounds back to me—including that one years ago at the Buddhist temple.

Later, in my hostel room, I realized that there was power in those chants. I recalled my hypotheses. All shamans see the universe as made from vibrations. All shamans perceive reality in an altered state of consciousness. And all shamans work with a sense of a higher power.

Somehow there was power beyond words contained in the vibrations of sound. Somehow this power altered the consciousness of both the shamans and the patient who was being healed. It had to do with how we use our language and how it conditions our minds. I remembered that in all of my shamanic experiences, the shaman chanted a sacred song. Clearly this meant that these songs were important in shamanic healing. Somehow these chants took me beyond space and time. Was I just wallowing in a metaphor? Or was there some connection between shamanic sounds and the ability to see beyond the walls of space and time? Could I relate this to my quantum physics training?

For some reason my mind, as if I were still participating in an ayahuasca ceremony, was behaving like some magical hologram that had been illuminated all at once. In it were records of all of my experiences dealing with sounds and vibrations. Normally my mind follows directions, and parts of my mind are illuminated in some order, just as parts of a hologram are illuminated by a laser beam. And just as a hologram is illuminated, a searchlight of my conscious "will" must shine through it before anything is witnessed.

But I had no will at this moment. Things were just flashing willy-nilly. For a moment the searchlight was out, and instead information was flashing through my mind in an order that I had not even perceived before. These flashes gave me an insight. I realized that my shamanic journey, my quest for truth, actually began many years before I conceived of writing this book. Suddenly, I was sitting in my office at my university. I had flashed back seventeen years of my life.

It was the spring of 1973. I had been back in the United States resuming my teaching duties for two years since my sabbatical. One day I received a letter from my English colleague and friend, John Hasted, cochairman of the department of physics at Birkbeck College, University of London, inviting me to spend three months in London working on a computer-animated film about physics.

Three months wasn't the world by any means, but it meant new people to talk to and it also meant the chance to work and talk with one of the most profoundly thinking physicists I knew of, Professor David Bohm, who cochaired the physics department at Birkbeck. Since money was also available for me to come the following summer as well, it occurred to me that I could spend a period of six consecutive months in London and return to the United States in December 1973. I discussed this possibility with my department chairman, and since a replacement for me was available, he was quite content with my being absent from the campus for one full term.

Consequently I telephoned Hasted and told him that I was coming to London in June and would stay through December.

In the first week of June, the day before I left, my ex-wife, Elaine, came to my campus house along with our children to say good-bye. My youngest son, Anthony, was only five years old. As I looked at him and then picked him up in my arms, I suddenly burst into tears. A flood of emotion overcame me and I saw that I wouldn't see him again for at least two years! How could I have known this? My grant from London was for only six months. SDSU was expecting me back the following spring of 1974.

When I arrived in London later that month, I soon began work on my film. I had to complete it in time for a conference in Belgrade, Yugoslavia, the next month. When I arrived in Belgrade, I met with another friend and colleague, Jean Durup from the University of Paris. I had met Jean in Paris in the late summer of 1971 as I returned from a round-the-world sabbatical. Durup told me that a professor from the University of Chicago who was expected to spend the academic year beginning that fall at the University of Paris had suddenly died of a heart attack in Chicago and therefore a post was open in Paris for a visiting professor. Would I accept it?

I thought of my youngest son, Anthony, and my feeling two months earlier returned. Now I knew why I wouldn't be returning to the United States for two years. Durup's offer carried a substantial salary, which, being tax-free, was a considerable sum for me—more than enough for me to stay in Europe for two years. I said "yes" nearly without thinking. Jean then told me that I would need to teach a course in quantum physics to graduate students, in French, but, other than that, my time was completely free to investigate whatever I wanted to.

Although I kept my post at Birkbeck College in London, Jean assured me that there would be no conflict and that I could begin my teaching duties in Paris in the winter.

In January of 1974, I looked for a place to live in Paris. A Parisian friend introduced me to a friend of his who was looking for a roommate to share his apartment with him. The apartment was in the center of Paris on Rue de Condé, right across the street from a set of flats once lived in by the Marquis de Sade. I took it.

By late February, my French lessons at the Alliance Française were proving fruitful. I must admit, though, understanding French was much more difficult for me than speaking it.

Learning to speak another language was more valuable to me than I had originally believed. In fact, I found out that one doesn't just speak French, one in a sense becomes French. By taking on the attitudes, facial expressions, and the persona of the French people around me I found that I could lose my American accent so well that, even though no Parisian was fooled a second that I was French, they had no idea where I was from.

This gave me a new insight into language: It was impossible to completely translate one language into another! Language contains more than just words. Words contain meanings that have been inculcated so deeply that no simple translation is ever totally adequate.

It was during this discovery that another step along the path of my quest was taken. I met a Qabalistic master, Carlo Suarés, who was to show me another insight—a new connection between my work in quantum physics and the spiritual insight marking my shamanic journey.

IN THE BEGINNING WAS THE WORD AND THE WORD WAS: "SATAN"

In early April, I had received a phone call from an old and dear friend. He called to tell me about a strange book he had received and read. He had already mailed it to me and wanted my opinion about it. The book was *The Cipher of Genesis*[1] and the author was Carlo Suarés, who just happened to be alive and, although not too well, living in Paris.

A few days later the package from Toben arrived. After reading the book and finding it "beyond me," I was nevertheless tempted to make contact with the author, and I did. Carlo and his wife, Nadine, invited me to their apartment.

When I arrived I felt as if I had returned to visit with my "cosmical grandparents." Carlo and Nadine were in their eighties. However, my meeting with Suarés was not friendly and cordial. I have the habit of sometimes being "too direct" in my inquiry. I told Carlo that I had read *Cipher* with interest but couldn't understand it very well. He appeared a little miffed at that and asked me what I did for a living. I told him that I was a physicist and had been using computers to make films. He professed no interest in computers and began to be quite bored at my intrusion into their seventh-story eagle's nest facing the tower of a World's Fair long forgotten.

Just as I was about to leave, I noticed that the room began to feel less tense. Suarés leaned back into his comfortable living room chair and remained silent. Suddenly I found myself talking about one of the images in his book. It was an image that long fascinated me. Indeed it is a primal archetype for the underworld. It was the image of Satan, the Devil. Suarés referred to Satan many times in his book. It was in one particular reference that I was fascinated. Satan was a code—a living code within us.

[1]Suarés, Carlo. *The Qabala Trilogy: The Cipher of Genesis; The Song of Songs; and The Sepher Yetsira*. Boston & London: Shambhala, 1985.

Satan is a Hebrew word. It is spelled in Hebrew, *seen-tayt-noon.*[2] In colloquial Hebrew it means the adversary or accuser as well as its English meaning. Each of its three letters is also a "word." They each in turn have meaning. According to the Qabala, *seen* represents a movement—a universal or cosmic motion—one that underlies the movement of all things in the universe. It is pictured by Qabalists as the "breath of God" returning from the material plane back to the infinity of God.

Tayt represents a primitive cellular structure. It could be a living cell, a memory "engram," a neural circuit, or any other analogy representing a basic structure exhibiting the cell-like behavior we understand. It is a feminine concept that tends to enclose energy repeating itself, as a bird builds a nest for egg laying. It is a self-referring concept, one that continues to build itself in its own image.

Finally, *noon* stands for a high principle operating within the cosmos. This is the principle of cosmic indeterminism.

When Suarés pointed these things out to me, I naturally was interested, since the principle of indeterminism is fundamental to quantum physics and I wondered if the Qabala had any relation to it.

I suddenly saw something in Satan. It came to me in a vision or picture. I saw a laser light beam entering the cell, and the cell was being blown apart producing many spherical waves of light, each containing a possible remnant of the cell. I told this to Suarés and he smiled. He said, "You are becoming a Qabalist."

Suarés had written that Satan represents the cells of our living selves held between the freeing movement of God's breath and the fearful possibility that anything can happen to it. As a result the cell tends to persist in living fear of life-death.

He pointed out to me that my vision of the bursting of the cell and the laser light beam entering it was thus a quite accurate description of the "action of the cosmic Satan." It literally disrupts the stasis of the cell, producing doubt and uncertainty. But this was not necessarily to be feared or prevented. Without the process of Satan entering the cell, no learning takes place. Nothing new can happen. We would be doomed to repeat all that we do.

But Satan guarantees nothing. Death of the cell can result with nothing being gained. Thus, with no guarantees, Suarés asked me if I would like to return to his home at some later date. As I was leaving, Nadine smiled at me and said that Carlo enjoyed our discussion in spite of the early

[2]In every use of Hebrew letters, I have followed Suarés's spelling of the letters. The usual spelling would be different.

difficulty we had in communicating with each other. Indeed she said Suarés found me a necessary "resistance" to him, one that would be vital to my understanding of the working of the universe. By resisting him, I was learning from him. By testing him, I was being tested. "Many have come before," she said, "and few leave with anything. They all shake their heads in nodding approval but leave with nothing in them."

Thus it was that I, a materialistic physicist with little or no spiritual training, especially in Hebrew, found myself a student of one of the foremost masters of Qabala. The mystical implications of the quantum world were to take on a startling new illumination. For the first time I was to experience a direct link between quantum physics and spirituality.

A few weeks later a number of friends, who all happened to want to visit with me from various parts of the world at one time, descended upon me. All of them were interested in meeting Suarés. Although his health was failing, he found the energy to discuss with us secrets, the kind of secrets I have always been excited about. One of these secrets was revealed on a bright, cold spring afternoon just as the sun dipped behind the Eiffel Tower.

We asked him how consciousness appeared in the universe. His answer was to affect my research and start me on a shamanic path. What he told me indicated that mind arose from a "movement" from here-now to there-then and a return again in a double flow from the present to the future and back. In other words, our minds weren't time-bound. Ultimately this revelation would help me to understand just how shamans do the things that they do.

According to his teaching, the universe was constructed from the vibrational sound patterns of three Hebrew letters, *aleph*, *mem*, and *sheen*, in an interplay of spirit, matter, and consciousness. He symbolized the spirit by the Hebrew letter *aleph*. Matter was symbolized by the Hebrew letter *mem*, and the wave movement of consciousness was symbolized by the Hebrew letter *sheen*. What we call "consciousness" consists of waves of information that move from spirit into matter and then back again into spirit. This flow of waves took place beyond time, in the sense that the whole action of that movement was instantaneous.

As he told us this, I had a picture in my mind. I saw God as a being blowing air into water making bubbles. I had the idea that the breath of God was the movement of consciousness, the bubbles were particles of matter appearing, and God was the spirit. Thus not only did God blow consciousness into the form of matter, but simultaneously this action was the awareness of those bubbles as matter.

Suarés had explained a basic mystery that I was just beginning to glimpse in quantum physics. I knew that according to my understanding of quantum physics, any conscious experience also results from a double movement of a wave action. This wave action also takes place beyond space and time. It is a wave of possibilities, what we call the quantum wave function, that moves from here-now to there-then and back again. I realized that he had explained in Qabalistic terms the very action of the quantum wave of possibilities that physicists calculate in order to predict the behavior of matter.

Suarés had made me aware that it was possible to see through time and space in a manner that at first seemed contrary to the laws of physics. I now suspected that shamans were able to "see" in this manner, and at the same time I realized that this ability may not exceed the laws of physics at all. They saw through space and time if they were somehow able to tap into the quantum wave function. I had no idea how they could do that, or even if this was possible. But there was another clue even if there was still a mystery.

He also had pointed out how important sound was in conditioning our experience of the world. He made it clear to me that sound had two different functions in the world. They were different and they were related to each other. The first was the formation of descriptive sounds, what we call ordinary words. These sounds signify ordinary experience, but the sounds are not the experience they point to. For example, the word "bath" is not a bath. The second was the formation of projective or sacred sounds. When these sounds were pronounced, a resonance of some kind would occur and the speaker would be in "tune" with the universe. Possibly these sacred sounds linked the enunciator with the cosmos in this double-flow action of quantum waves. I was guessing. However, I felt that sacred sounds existed and the mere pronunciation of them could invoke the sacred experience that they pointed to. Later, I was able to prove my suspicion. These sounds were able to change matter.

My memories of Suarés and the image of the chanting shaman in the film were overlapping. I now suspected that sounds and words can be used to heal or harm—a primary lesson from the shamanic world. And I was beginning to see how this all connected up with quantum physics. I hadn't realized until now, in Lima, that Suarés had explained to me just how consciousness worked in terms of quantum mechanics. Perhaps the link between sounds and words and quantum physics existed in the observer of these vibrational patterns.

The shaman's chant continued. And as I watched the young bearded scientist sitting in the shaman's hut next to Nexy, more images of myself as a young professor filled my mind. It was my next lesson in the power of words and sounds and I was back in Paris again. And it was after this encounter that I began to realize that illness and the words we choose to describe that illness were more deeply connected than I had ever suspected.

THE "SHAMAN" IN THE GRAY FLANNEL SUIT

By late spring, I had been totally absorbed with Suarés's teaching. He had made it clear to me that our language played a major role in conditioning our perception of the world around us, including ourselves. In fact I began to think that we couldn't really perceive the world unless we had some form of language to think about it.

Although I was occupied with teaching at the university, I was fully enjoying the Parisian ambience. The spring was turning into summer. The trees in Paris were in bloom, and the air had a wonderful new warmth to it. The Jardin Luxembourg was beginning to fill with people, and romance was alive and well and living in Paris.

By then, I had finished my teaching duties in Paris, so I returned to London, a trip I had been making quite frequently during the past year, since I still held my honorary fellowship at the University of London, even while I was teaching in Paris.

It was mid-May and a group of English "New Agers" decided to sponsor a "New Age" conference in London called, appropriately, *The May Lectures*. A number of luminaries of the New Age were coming to speak at the conference. Lyall Watson, Andrija Puharich, Werner Erhard, Fritjof Capra, Ida Rolf, and many, many others were invited speakers. I naturally wanted to attend and managed to get a ticket for the lectures.

One of the speakers was presenting a shortened six-hour workshop experience of a two-weekend package seminar his corporation was giving in the United States. I had never attended a "New Age" workshop before and wondered just what people did for such a long time sitting together. When I arrived in the room, I was surprised to find several hundred people in attendance. I was even more surprised to see that the workshop leader was slickly dressed in a suit with all the spiff and polish you can imagine. He appeared to be a salesman at a smart car salon rather than a leader in the blossoming New Age movement. I wondered just what he was going to do.

He welcomed everyone to the seminar, saying what a rare treat it was for him to lead this workshop before peoples of Great Britain. He proceeded to compliment everyone, talking about the war years, and the reconstruction programs going on there, and the splendid spirit he found in the British Isles. I noticed, of course, that as he talked the people in the room, who seemed to me at first to be a little uptight, especially after seeing this slickly dressed "American salesman," visibly appeared to relax.

But then he said something that amazed me. "Well that's the last nice thing you are going to hear me say. From now I'm not going to be nice. We are going to explore just what it is that keeps you from reaching your full potential, and to do this won't be nice," he said. He then proceeded to say things that to me were not offensive in the least, but to a British audience, they were as devastating as bombshells. As he went on, people were shouting, "We won't do this." People began to stand and shake their fists. Some proceeded to leave with their noses completely bent out of joint. A few tried to take over the room, leading snakelike Conga lines, chanting and singing old English war songs and the like, as they marched around the room.

In one hour this amazing man had transformed an audience of nearly stoic Britishers to a mob of unruly and nearly crazed animals. And yet, as surprising as it was, he had never lost control of the group. One sharp retort from him and the whole room, as if hypnotized, would suddenly sit down and shut up.

This performance was better than any stage magician's I had ever seen. It was the first time that I was able to witness a shamanic activity. I call this *voice magic* in action. Later, after everyone had quieted down, and he had begun a *process* to enable people to learn how to visualize what they wanted from their lives, he surprisingly announced that he was interested in meeting physicists, and that if there was any science at all that could explain the work he was doing, it would be physics. He then announced a break of about fifteen minutes. He walked to the back of the room and out the rear doors into the foyer facing onto the seminar room.

I naturally was curious to meet him, but I never thought it would be possible. I expected to see him surrounded by his aides, protecting him from the apparent wrath of the attendees. I left my seat and quickly made my way through the milling crowd to the rear door.

I found him standing next to a wall, completely alone. No one was approaching him, no one at all. I walked up to him. "I noticed that you said you were interested in meeting physicists. I am a physicist and I have been getting involved in the study of consciousness. My name is Fred Wolf," I said. He looked at me quite interested and focused. "Pleased to meet you. I'm Werner Erhard," he said.

I told him that I was completely amazed at what he had accomplished in that room and I wondered about his interest in physics. Instead of answering me, he proceeded to ask me a number of rather specific questions indicating that his interest in physics was more than just a passing one. As we continued to talk, the break time came to an end, and Werner then invited me to come to an all-day session, for speakers only, that was being held at a small college to the west and just outside of London, the next day.

The next morning I found myself attending a small group session of the workshop leaders. The discussion was on transforming planetary consciousness—no small topic even for world leaders to get into. As the discussion continued, I noticed that a lot of assumptions we usually hold about the world, and the problems inherent in it, were not due to physical constraints, but to the mental constraints we all carry with us. In other words, the world and all of its successes and failures, its trade agreements, its border invasions and war states, its military budgets and welfare programs, its languages and acceptance or nonacceptance of others who don't speak those languages, are constructs of our thoughts and our belief systems. Even money is such a construct.

These leaders in the consciousness movement clearly had a new vision, and my mind, always accepting and believing that real tangible forces, such as lack of material resources, drought and famine, were the real causes of human suffering, was being blown that sunny morning at a small English school just outside London.

I realized that the world, as we saw it, was just a question of agreement of those that participated in it. Having studied quantum physics, and having realized that there is such a thing as the observer effect—where, as you'll recall, the choice of an observer to measure a particular physical property actually creates it or brings the state of the property into existence merely by observing it—I wondered if the world, too, was just a construct of our own thoughts. Not just the world of consciousness and thought, but even the real physical world that Blackstone and other clever magicians were seemingly capable of manipulating before our bewildered eyes.

The truth of this was dawning on me. There were no limits except those limits that we had imposed on ourselves. The mythic world of ideas and visions and the "real" world of materiality and people were now overlapping in my mind.

Suddenly, I found myself standing from my seat, and saying:

"I just got back from Paris where I was studying with a Qabalistic master, Carlo Suarés. Suarés's teaching seems particularly relevant to this discussion. One of the Qabalistic words that has special meaning for us is the word that is spelled *dallet-vav-dallet* (d-v-d). *Dallet* is the symbol for resistance and at the same time the symbol for response. *Vav* means the act of fertilization. So *dallet-vav-dallet* means resistance fertilizing resistance, or if you will, it means response fertilizing response. Without this action, which by the way is pronounced 'dovid,' from which our word David comes, there is no possibility for any transformation to occur. We each resist the other. We need to learn to transform that resistance into a yielding response rather than a wall-like reflection. This word, dovid, by the way, means in Hebrew 'lover.' Thus without the 'lover' there is no transformation possible."

When I had finished I noticed that all of the leaders' eyes were on me. Apparently, and seemingly unwittingly and yet profoundly, I had added a new dimension to the group. I remember Dr. Puharich particularly saying, "Thank you, Rabbi." Now I had never considered that I was anything like a rabbi, so I was taken aback by this. "Why have I said this?" I wondered. It was much, much later that I was told by a "psychic" with visionary power, that I had two spiritual allies working with me. One of them was an old rabbi, and the other was a sphere of spiritual power with no discernible humanlike features.

The images of myself as an "old rabbi" began to overlap on the screen with the face of the young hero. Tears were entering my eyes as I remembered again that I was now older. My beard had grayed and I was still searching for a truth that I was just now realizing.

I realized that all shamans deal with a sense of the spirit world. They told me that they all had spirit guides that spoke to them and told them insights into healing. Shamans would invoke these spirits by chanting sacred songs. Perhaps there was a connection between these chants and sacred use of language used by the Qabalists.

My experiences with Suarés, Werner Erhard, and the group of "New Age" leaders convinced me that words had a special power if used in a certain manner. The word "dovid," dallet-vav-dallet, response fertilizing response in a resonance of vibrational energy, meaning "lover," made much sense to me now, after my investigations with shamans, in a way I could not have suspected at that time.

Indeed it was one of the secrets of healing that they taught me. The word "dovid" symbolized the resonant transfer of vibrational energy from one person to another. Could that same principle work in the human body? It then seemed clear to me that healing was a transfer of vibrational energy in the body. When a body is ill, parts of it are out of harmony with the rest of it. All organs and cells vibrate. Perhaps when an organ is diseased, it is no longer receiving vibrational energy from the rest of the body. It was vibrating at the wrong frequency, out of harmony with what nurtures it.

But how did that happen? And how could it return in resonance with the healing parts of the body? I then remembered. Suarés had taught me that there was a sacred use of language that, in our hurry to develop a descriptive language, we had lost. The Qabalists believed that the universe was created from sound. Thus by reciting sacred sounds, changes or transformations of matter could take place.

Healing could simply consist of reinvoking those sacred sounds in the body. In other words, by singing or reciting the correct sounds, various parts of the body, out of harmony, could be brought back into harmony. Perhaps this is what the shamans do when they chant. I was still confused about this.

Then I recalled something from my early learning. Shamans invoked changes in nature by calling for them to appear. They didn't just call out, they sang the words. A word, when spoken in a certain way, invoked the thing spoken and not just a symbol of the thing spoken. Perhaps this is what Suarés had in mind when he told me about the sacred use of language.

For example, in the shamanic tradition the chanting of the word "wolf" was an invoking of the wolf itself. Thus if one sang the name of a wolf in a sacred manner, a wolf would appear. The Qabalists, the ancient Hebrews, believed this too. In other words, a sacred sound was connected to a real object, and the object could be invoked by the sound or, if the shaman was powerful enough, even the thought of the object.

But for the shamans, invoking the wolf also meant identifying with the wolf that lives in each of us. Shamans somehow knew the wolf's vibrational pattern. They were able to resonate with wolves and other animals. Each animal is represented by a totem spirit of that animal, and that means that we are brothers and sisters to all living kind.

This rings true of the idea taken from quantum physics, namely that there is a fundamental interconnectedness, a web of meaning hooking everything together. I realized that shamans see the universe in a more interconnected way than any way envisioned by mechanical models such as Newtonian mechanics or Galilean relativity or for that matter in any of the models of the universe that were pre-quantum or pre-new physics. Shamans don't just see things interacting with things producing cause and effect relationships, but they see a spider's web of interconnectedness which is very close to the kind of interconnectedness that we see is present in quantum models.

In physics, we call this nonlocality. It means that actions taking place here can instantly affect actions taking place there. In 1964, an Irish physicist named John Stewart Bell[3] made a set of observations that showed within what limits one could experimentally determine the presence or

[3]See reference to Bell in the bibliography.

absence of this connection. In other words, suppose there were a certain set of experiments we could carry out with two objects that interacted and then separated, would we be able to tell the difference between whether the second object actually had a particular characteristic before the measurement or had to assume that characteristic as a result of the measurement on the first object?

He showed that the second object definitely had to assume the characteristic as a result of the measurement on the first object, and he established a precise physical limit as to when this circumstance had to take place.

In my mind the two objects were the sacred words and the object invoked by those words. In this sense the modern view of physics and the ancient view of the shaman have a very similar context. By speaking the word "wolf" (corresponding to the measurement performed on object one), the wolf appears at the door (the second object takes on a physical characteristic), and we take on very briefly, if at all, the resonant energy of the hunter, the wolf.

With quantum mechanics rediscovering the ancient view through a very peculiar way of proceeding, by investigating the very minutest aspects of matter, some quantum physicists came to the conclusion that, one way or another, matter could not exist without some consciousness to perceive it.

Now the word consciousness is a great "bugaboo." Not every physicist today would admit that consciousness plays a role in the physical world or say that consciousness was even a part of physics. They would use another word to describe the action taken when an observation occurs. They would say registration, or measurement, or recognition of, or preparation of, a state. They would use a lot of different words to indicate that somehow a pattern of possibility-recognition has been reduced from a multitude of possibilities to a single precise exact result.

How that is accomplished is still subject to a great deal of debate in physics. But there is not a physicist around that would deny the fact that without a recognition of that pattern or a precise measurement of some aspect of that pattern, that pattern itself could never be perceived.

The perception of that pattern and the appearance of an object in the universe with the objective qualities that are contained in the pattern are the same thing. This aspect of the observer effect is extremely important. Consciousness and the material world are connected. Just how an observer chooses to observe not only affects him or her but also affects the object being observed. It connects the observer and the observed in a meaningful way.

That connection was realized whenever a shaman invoked a sacred sound. He was resonating with totem spirits within himself. In this way he was able to heal not only himself but anyone else who suffered from a similar illness by being disconnected from that totem animal spirit. But what about we ordinary mortals who simply utter whatever we wish with no regard to the sacred meaning of those sounds? All we could do was invoke random noise on the circuit of human discourse. Perhaps it was this simple fact that cut us off from our natural healing abilities.

This I realized as I watched the film and listened once again to the soothing songs of the shaman playing on the soundtrack as he healed the woman in the film. I wondered again, could we reconnect with our sacred heritage of sound? Again pictures flashed through my mind. I was once more back in England, but this time it was 1988.

It was then that the full power of words hit me. I had journeyed back to England as part of my research on this book to meet with two Englishmen who were practicing Anglo-Saxon shamans who had been researching and practicing the *sounds and ways of wyrd*.

THE SOUND OF WYRD

In August of 1988, I made my way by car to the south of England to meet with Chris Hall. Chris is a noted researcher into the ancient history of Europe and has had a number of extraordinary experiences dealing with shamans. From his reputation, I suspected that Chris must have had some shamanic power himself.

In the evening, after I had found a hotel in Brighton, Chris picked me up in his Austin Mini and we began to talk as we drove around looking for a convenient "pub." I asked him about the meaning of the word "wyrd."

He told me that the word "wyrd" comes from the Anglo-Saxon language. The literal translation of "wyrd" means "to become, to be, that which is." According to ancient traditions, the Anglo-Saxons had a whole range of gods that possessed superhuman powers.

These heathens were quite sophisticated. They believed in three abstractions—concepts which were greater than the gods and which even the gods were subject to. These were time, causality, and fate. Fate was the major one.

When the Christians came into power, everything that was heathen was instantly bad. So every meaning got reversed, exactly 180 degrees. This reversal has affected the everyday English language. Today, something which is "weird" is strange. But in those olden days something that was wyrd was normal. This is similar to the way American blacks use the word "bad." "It's really bad, man," means that it is really good.

So, according to Chris, something that is really wyrd is something that has a necessity, an "isness" that is so palpable as to be undeniable. Wyrd is that which is and that which will be. It is the inexorable unfolding of the universe. The heathens pictured it as a multidimensional cosmic web. In it, everything had matter, spirit, and soul. Everything was linked. So when I hit a table, the reverberations go throughout the universe. No matter what you do, it's part of wyrd.

This meant that in ancient traditional belief systems like the Anglo-Saxons', vibration and vibrational patterns were extremely important. The Anglo-Saxons used symbols for that importance called *runes*. Runes are characters of an alphabet probably derived from Latin and Greek and used by Germanic peoples from about the third to the thirteenth century. Like Hebrew letters, each rune also represents a word. Runes are taken to have sacred and divinatory powers, much like the Hebrew letters in the Qabala. I explored with him the nature of sounds and words as depicted in runes. I told him about my work with the sound of the Hebrew letters. Chris confirmed my thoughts, "Runes are an attempt to do what you have done with the Hebrew letters, to express the universe in sound."

Soon we made our way to a noisy bar. While quaffing a round of ale, Chris began to talk once again about the early Anglo-Saxons and their connection with shamans, telling me that the early shamans spoke in a very special manner that he had been fortunate to learn.

SHAMANIC POETRY

He told me of having been invited by a group of people called the English Companions, a group of Saxonists, to give a talk in the Tower of London on the early Anglo-Saxon world view. Chris gave his talk in a very special manner known in the Anglo-Saxon times as *Goldar*, which means magic. Goldar reciters held their audiences "spellbound," which incidentally is where this term comes from.

Chris reminded me that words held magical power. Goldar was a shamanic technique that he called shamanic delivery. You begin by speaking slowly and pulling people in. By the end of the talk he was speaking in what he calls "note form."

He told me, "Yet everyone was hearing complete sentences. I checked with the audience afterward to see that this was true. They were getting my meaning, but I was just talking in short staccato images. I went faster and faster and in the end it was like gobbledegook. But they gave me a standing ovation."

I remembered a film called *Zardoz*, produced by John Boorman in Ireland years ago. In the film a group of people living in the future possessed incredible psychic powers enabling them to engage in a form of storytelling in which the storyteller spoke in "notes," as Chris had done. Consequently only partial sentences or, for that matter, partial words or sounds were uttered by the storyteller, and yet everyone listening understood the tale being told.

Chris continued, "Just before I went on, the Armorer of the Tower of London came over to me and said, 'May I tape-record your lecture. It's a standard practice and we use it in our education programs for schoolchildren.' I thought, 'Oh, my lord.' I felt that this was going to be unintelligible on the tape. I said that he could try. I don't know what made me say that. Anyway, he went into his recording room and pushed his buttons. Although everything seemed to be working, in the end nothing was recorded. Actually this was not quite true. You can hear two words that I uttered, 'Thank you' at the end of my talk."

We then moved from the bar and went back to the car. As we began driving, I wasn't sure where we were headed. We finally ended up in another bar called the Prince Albert. Staying long enough for another round, we then left.

WORDS AND WYRD

Soon I realized where we were headed. We were going to meet another shaman and friend of Chris's named Richard. As we entered the gate to his home, I noticed a number of dogs surrounding the fenced cottage. I thought that this was especially appropriate since Richard was supposed to be a shaman. I entered the gates somewhat reluctantly since the dogs were snarling. Chris told me not to mind the dogs, so I didn't. Richard, a tall lanky blondish fellow, who looked quite a bit like a Nordic Viking, bid us welcome. Soon we sat down in his small living room.

Richard appeared surprisingly gentle for such a big fellow. In comparison Chris was quite a bit smaller, perhaps about five foot seven or so. Richard must have been at least six foot three.

After some tea, Richard came up to me and asked me to hold a drum against my sternum. At first I experimented, trying to find the right spot. Richard placed it directly on my bone. He said it was best if I held it so one drumhead rested against my chest. While holding it, I struck the drum, first with the soft end of the striker and then the hard end. I felt the drum vibration all through my body. It made me shiver and I felt a ringing throughout my body. Even though the drumbeat had a low tone, I felt it ringing in my ears.

Richard pointed out that this was a tuned drum, quite particularly tuned to resonances in the body cavities. The sound felt like a little flicker of electrical energy going up my body. He said, "You can see how such a thing could become melodically hypnotic. You can create a state of euphoria."

As Richard spoke, I experimented with the drum and the striker, turning it and hitting the drum with different parts of the striker. As I did so, the tone raised in frequency. It almost sounded like an alarm bell. I then felt a definite shift in my focus of attention. It wasn't that I fell into a swoon or anything like that. I just was more attuned than I had felt before I held the drum.

I began to think. Could this be an important link to the relationship between physics and shamanic practice? Perhaps vibrational patterns set up in a sick human body, by the shamans when dealing with an ill subject, caused the subject to be cured or rid of any spirit infestation.

These two shamans were telling me a secret. As they moved around the small cottage showing me one vibrational tool after another, as we spoke of words and meaning, they were instilling in me a key to shamanic healing. A simple truth. It was all a matter of vibration.

According to principles of quantum physics, the universe is made of waves of probability—vibrational patterns that ebb and flow with particular frequencies. This pattern reminded me of Chris's statement about the web of wyrd.

Each frequency relates to the mass of an object. Alter that frequency and you in effect change the mass of the object. By such a change, even a very subtle change, the object could be slightly displaced from a particular location and literally moved in a quantum jumping manner to a new locale. What was out of resonance would come back in resonance.

I knew from my own work that nerve cell vesicles—tiny messenger molecules that are released in the synaptic gap between two nerve cells—operate probabilistically. That means that they are subject to quantum physical laws of motion and that they are affected by these waves of probability. I could imagine the shaman inducing such changes by causing the body cavities to vibrate resonantly. Such a vibration would shift the organs in a subtle manner and send new messages down the nervous pathways. These messages, carrying the information of the shifted organs, would put the diseased organ in touch with the rest of the body. And then, somehow, the body would heal the organ itself. The question was how?

There had to be some form of resonance. Perhaps by using sound in a certain manner, that resonance would occur. But somehow the person being healed had to be able to allow a resonance to occur. Again, but how?

Our conversation continued through the night. My mind was in a whirl.

I had absorbed all I was going to that night. I said good-bye to Richard, and Chris and I left. As we drove back to my hotel, I asked Chris, "What is Richard's last name?" He told me it was Dufton. "His full name is from the Baronetcies of the Duftons."

"Is your first name Christopher?"

"Yes, it's quite an interesting name for a student of heathenism. It means follower of Christ."

Later, after the movie in Lima, I recalled my thoughts. Inherent in the shamanic world view that Richard and Chris told me about was the notion that all things were vibrationally connected. As I mentioned earlier, ancient Anglo-Saxons saw this connection in terms of a spider's web—the so-called web of wyrd.

In terms of the "web of wyrd," the whole is composed of parts that are connected just as dew drops of water are held by a spider's web. One drop vibrates in the breeze and the whole web responds. Every drop feels the breeze. But what was the web that connected us together? I felt that it had something to do with consciousness, but I wasn't sure. I also suspected that this vibration could heal.

Jorge Gonzalez, the Peruvian shaman, also told me that everything was vibration. Candace Lienhart, an American shamanka (the female word for shaman) I had visited in Arizona last year, had also told me that when she healed anyone she used her body as an antenna picking up on organs that were not vibrating correctly.

But what was this vibration? During my first shamanic ceremony with Jorge (described in the next chapter), I had felt it, and yet I still didn't know what it was or how it manifested.

Just a few nights ago, when I was participating in my third ayahuasca ceremony which took place in the Amazon jungle, Don Solon, the shaman who had taught Jorge, had made the floor of the jungle hut vibrate in some mysterious manner that I still cannot figure out.

It was clear to me that the first hypothesis was correct. Shamans see the universe as made from vibrations. But how were those vibrations connected to the whole universe? My experiences with sacred sounds and chants made part of this clear to me. Richard's drum made me feel quite strange when he struck it next to my chest. Somehow my emotions were connected to the drumhead. Somehow the striking of the drumhead altered my consciousness. But how? And why had I felt Jorge's vibration but been unable to fully explain it to myself?

There was a connection between sound and the body. Suarés's teaching coupled with what I had experienced with the New Agers indicated that the body had several vibrational modes present. Illness or stress would alter those modes. When a proper sound was presented to the body, the body would attune to that sound, and just as the metaphor suggested, the body would come into harmony.

I was struggling with the shaman's higher power, the seventh hypothesis. Then, suddenly, the idea of resonance began to occur in my head. Somehow there was a resonant interaction between the drumhead and my body, between Don Solon and the floor of the jungle hut, and between Jorge and me. Somehow I had felt connected to a fly in Katmandu. And yet I had only sensed those vibrations under very special circumstances. I was in the presence of shamanic influence. In my normal waking condition, I hadn't felt anything like that before. I now needed to find out what shamans had done to me to create my condition of accepted resonance vibrations with the universe they had created around me. I then realized what it was. They had initiated me. They had made me a shaman's apprentice.

4

Rituals and Initiations

The movie shaman's song continued. My mind was flashing back through time, seemingly beyond my own control. And now, Miguel and Nexy were undergoing their shamanic experience. They were experiencing an initiation ceremony, one that unbeknownst to them was preparing them for shamanhood. The shaman in the movie was rocking back and forth as he chanted his sacred song, his *icaro.*[1] I remembered the power of words and sounds. They were just vibrations.

Later that evening I realized there was more to the story. Could that be the essential clue to the healing practices of shamanhood? Was it all just good and bad vibrations? What had these vibrations to do with initiation? I recalled my hypotheses. My first hypothesis stated that shamans see the universe as made from vibrations. My second hypothesis said that shamans see the world in terms of myths and visions that at first seemed contrary to the laws of physics. I wanted to know how these myths were created. I wanted to know how shamans were able to seemingly alter the laws of physics. Perhaps they didn't. Perhaps it only seemed that they altered the laws of physics. My fourth hypothesis also seemed to be relevant here: Shamans will use any device to alter a patient's belief about reality. And my fifth hypothesis entered my mind: Shamans choose what is physically meaningful and see all events as universally connected.

[1]An icaro is a sacred healing song chanted by a shaman.

Now, there are two important events in the consciousness strand of the web between a shaman and the people who need the shaman's healing energies. Both events are observed in a testing ceremony. The first event is when the shaman convinces himself or herself that he or she has genuine shamanic power. Shamans are ordinary people, too. They must test themselves every time they conduct a healing ceremony.

The second event takes place when the person who is expecting to be healed is convinced that the shaman has the power to do so. Thus there are two event-points of consciousness required in order that a shaman be real. It's a two-level "confidence game" requiring that both the healer and the healee believe in that power.

Thus in the world of the shaman and patient, a particular initial event (something the shaman does to convince himself of his power) and a particular final event (something the patient sees that convinces him the shaman has the healing power) must be meaningful for both partners in the dance of healing.

All of these hypotheses pointed to the observer effect. The patient needed to be prepared. According to quantum rules, the rules that govern the behavior of matter, an observer must prepare the state of the system he or she hopes to observe. It was only logical that the patient, the person needing healing, needed to have his or her own body and mind prepared in a special way.

Now, in a physics laboratory, that meant preparing the experiment in a very careful manner. If special pains were not taken, the experiment was bound to fail or, at best, give poor results. Everything had to be carefully isolated so that spurious effects would be diminished. I envisioned the shaman as a kind of physicist. Could something analogous be happening when a patient interacts with a shaman? Perhaps the shaman prepares the patient just as carefully as the physicist prepares his experiment. Curing a patient was like getting the desired result in an experiment. Thus a shaman puts the patient through a ceremony, an initiation period. This must be done before the cure actually takes place.

I then remembered my train journey from Machu Picchu, just a week ago, when I recalled my thoughts about shamanic vibrations and the connection they had with the rituals I had experienced.

The sun was setting as the train from Machu Picchu made its way toward Cuzco. The constant rocking of the train and the repetitive clickety-clack sound it made as it went along the tracks started me thinking again. I noticed that my body was rocking back and forth with each turn and vibration of the train. I noticed that if I fought the vibration my ride became uncomfortable, but if I moved my body in harmony with the train, I felt the energy of the train in my body.

My theory was that between a shaman and a patient there had to be a similar resonant interaction. This interaction was something like putting two resonant systems together. The key was the ability of the shaman to tune to my frequency of vibration and then absorb the illness from me. The better I was prepared for the experience, the more the shaman could absorb.

I pictured two pendulums hanging from a ceiling. Each pendulum was able to swing back and forth freely. But because they had different lengths, the shorter one swung faster than the other. They weren't in tune. I pictured connecting the two of them with a weak spring attached to the rods holding the pendulum bobs. If you set one of them swinging, the other would start to pick up the energy. But if the two pendulums had different lengths, the coupling would be poor and little energy would transfer from one to the other.

But if you adjusted the length of the resting pendulum to match the length of the swinging one, the active pendulum would slow down and the passive one would begin to swing with the same energy that the active one had. Then the active one would become passive.

The key was the adjustment of the length of the passive pendulum. I pictured the passive pendulum as the shaman and the active one as the patient. Somehow the shaman was able to "adjust his length" so that he picked up on the "ill" vibrations of the patient.

Or maybe not. Perhaps the active pendulum's length was changed to match the passive one. Perhaps it was the patient who had to change. He somehow had to tune to the shaman's frequency. This seemed more plausible to me. It also explained to me why shamans cannot heal everyone. They don't change. They somehow convince the patient to change. This meant that a patient had to undergo an initiation.

But now back in the theater, suddenly the scene had changed. It was a flashback when Miguel first came to Iquitos. Miguel was at a bar where Nexy worked as a prostitute. Jorge, the Peruvian shaman I had met in the United States and then later in the jungle, was in the scene, playing his part of a playboy. Surprised, I laughed. Which Jorge was the "real" Jorge? And then my mind flashed back. I recalled again what had happened to me a few months earlier and how I had felt that I was on a destined path just as I was feeling this watching the film. It was the time of my first meeting with this Peruvian shaman/actor, Jorge Gonzalez-Ramirez, in March of 1989. The meeting occurred in Santa Fe, New Mexico.

. . .

Jorge, when he is not acting in movies, is a professor of education at the National University of San Martin in Tarapoto, Peru. He is also a full-fledged ayahuasca shaman and healer, having been trained by shamans in the Peruvian jungle. The prospect of meeting with him was quite exciting to me, since he appeared to be well versed in both the world of the Western-educated scientist and the world of the mystical shaman. I wondered how my visit with him would turn out.

I didn't know it then, but Jorge was also taking me through an initiatory ceremony and at the same time healing me.

A JOURNEY INTO THE UNCONSCIOUS

My wife, Judith, and I met Jorge at the home of Jorge's friend who was sponsoring his visit to the United States. I hadn't realized then that Judith and I were starting an adventure with Jorge that would take us both well into the night hours and, with the ingestion of ayahuasca, even into a mind trip to the Peruvian jungle and beyond.

I told Jorge a little about my background. Jorge listened very attentively to both myself and to Mike, a shaman-apprentice to Jorge, who was providing the translation of my words into Spanish. I gathered that Jorge was not a wordy fellow, but what he said needed to be paid attention to.

I told Jorge about the observer effect and how observation alters matter. This was different from elementary classical physics where an object is solid and appears so regardless of how you look at it. In the quantum world view, I said, everything is like waves of probabilities rather than solid objects.

I asked, "Does that concept have anything to do with what you do in your work as a shaman?"

Jorge responded, "Shamans not only see, they also feel. So their action of seeing is a property not only of the eyes but of all the senses. We pick up the physical and/or spiritual or emotional problems that people might have."

What interested me here was Jorge's statement that feelings and sensations both affected him when dealing with healing. I suspected that Jorge used feelings and sensations to diagnose illness and that he dealt with the world in terms of feelings much more than modern scientists did. We perceived the world in terms of our sensations only. Perhaps reality was also shaped by feelings, even though physicists fail to deal with this.

I was beginning to get excited about this discovery. It had never occurred to me that there were two separate realities in the world, as it is experienced. One was based on our thoughts and our sensations. The other was based on our feelings and our intuitions. This had made sense to me; I just hadn't realized it before.

According to the Jungian model of psychology, there are four separate psychological functions. Thought and feeling represent two complementary functions. So do intuitions and sensations. In a previous book and in a journal paper[2] that I had written earlier, I conceived of a model of the Jungian functions based on quantum physics. Accordingly, a thought is complementary to a feeling in the same manner that the position and the momentum of a particle are complementary to each other.

Let me explain this. Quantum physics indicates that it is not possible to know simultaneously both the location and the momentum of a particle. A measurement of one necessarily disturbs the other. Thus a measurement of the momentum of a particle causes the position of it to become indeterminant, and vice versa.

In a similar manner it is not possible to simultaneously hold a thought and a feeling about something. One causes the other to become indeterminant. Jung had noticed that people's personalities developed according to the maturation of one function at the expense of the other. Thus a person who had the feeling function well developed would not be a mature thinker. And similarly for developed thinkers; they would not be mature feelers. One only has to think of the stereotyped scientist who, although quite intelligent, has the emotional feelings of a preteenager. The image of Dr. Strangelove in Stanley Kubrick's movie came to my mind.

Modern science has come a long way following the thinking function. Many scientists probably fail to see that there exists another scientifically valid way of evaluating and perceiving the world. This is particularly true, I believe, in the field of medicine. Scientists (and I include all medical practitioners as scientists) fail to see that when they proceed with a scientific analysis using their thinking functions they are actually entering into an altered state of consciousness and changing the way that they perceive the world—whether it is an electron or a patient in a doctor's office. They are putting on scientific glasses, so to speak, and changing the observed by so doing.

Jorge was no different when he looked at a patient—only he was putting on his "feeling-intuitive" glasses. By looking at the patient in this manner he, too, was altering the patient. It was the observer effect in action. I then asked how Jorge changed his perception when he entered a shamanic state.

[2]Wolf, Fred Alan. *Star*Wave: Mind, Consciousness, and Quantum Physics.* New York: Macmillan, 1984; *and* "The Quantum Physics of Consciousness: Towards a New Psychology." *Integrative Psychology.* Vol. 3, pp. 236–247, 1985.

He told me that he and the other shamans he has worked with or studied under empower themselves with magic potions through purges.

I asked him if that meant that at the moment it would be difficult to do this without the magic potion.

Mike answered that it would depend on the shaman and that it was a continuum. Mike said, "In fact Jorge does it both ways. It is much more powerful with the magic potions, and it is still powerful and it still works, and we still do healings, without the magic potions. You do enter into a trance just with the chanting and the ritual. But the magic potions are a quantum leap, if you will, as well. It is just more reliable with the purge." Jorge nodded his head in approval of Mike's words.

Jorge said, "With the purge we empower it, but also without it." He then invited Judith and me to stay for the evening session so that I could see directly just how he healed. I was reluctant, but I wanted to undergo a shamanic healing ritual, so Judith and I remained.

We then took a break and had some tea. Later, I asked him to explain what happens when you journey using ayahuasca.

He said, "With ayahuasca one goes to search in the subconscious world. It is a world of life. It is the world of recuperating normal life."

He told me that first you feel a dizziness that becomes more and more intense. Then there is a separation between the conscious and the unconscious world. This separation permits us to analyze, in a conscious manner, our whole unconscious history that happens to be guarded, kept someplace. And that is always the source of all of our pains and all of our misadventures.

I wondered if under ayahuasca you would have certain revelations that occurred as thoughts or as pictures.

Jorge told me both can occur. It can be very explicit or very symbolic. This is why the shaman needs to guide the self-analysis that each person is doing. This is why people who have common lives are separated in a session. Jorge said, "We don't let them sit next to each other. Each electromagnetic field is one, and we separate them. The fact of the act of making conscious that which was unconscious permits us to face our own selves. Not only face ourselves, but make a union or pact with ourselves. It permits us to look at something we are doing and see that it is bad and agree not to do it."

Jorge explained that ayahuasca could be a new therapy for drug addicts or alcoholics. The idea excited me. But it wasn't until later, after I had taken ayahuasca myself, that I knew what Jorge was talking about. By giving addicts ayahuasca, each addict would be able to "see" into his or her own addiction, because ayahuasca unlocks the unconscious mind. It was Jorge's insight that once an addict saw the root cause of the addiction, the craving for the addicting substance would vanish.

As I understood it, shamans feel that illnesses are not only somatic but also psychological. That is what Western medical people forget. They have divided the body into many separate parts, and they get lost in all those parts.

I looked at Jorge and then at Mike. I told them that much of this made sense to me. In my conversations with other shamans and other healers I had found many striking similarities. It made me realize that shamanic practice was a science. Even with the Anglo-Saxon shamans I talked with, or the American Indian shamans, or anywhere else, there is a great similarity in the practice.

Soon the sun began to set, signaling the beginning of the evening session. We then began to clear the living room of any furniture in preparation for the ritual. Jorge, a bit oblivious of all this, sat down in the corner of the room and began to meditate. While our host was in the kitchen, Mike and I moved the furniture out of the way and turned the living room into an empty, nearly nondescript space. We also covered the windows and then waited for the other people who were to participate in the ceremony to come. At 7:00 P.M., they began to arrive. It was obvious that many of them knew each other and also knew Jorge. Judith and I were, apparently, the only newcomers.

By 7:30, about twenty people, including Judith and me, filled the room and began taking our places on the floor, sitting on pillows as close to the walls of the room as we could manage. Some had brought blankets. Some were wearing sweat clothes. It looked like an old-fashioned pajama party but this time with adults rather than children.

Then the lights of the room were dimmed. Mike joined Jorge at the front of the room. A ritualistic altar had been prepared in front of Jorge. People placed various objects and possessions to be blessed in front of him. I put down copies of my books.

Then I noticed the bottle containing the ayahuasca at Jorge's feet. It appeared to be filled with a thick, viscous brown liquid, not very appetizing to the eyes, to say the least. The room had grown quiet, and then Jorge sat on the floor with his legs crossed in front of us, and the ceremony began. The lights were made quite dim. Jorge spoke. He told us that we were going on a journey into the Peruvian jungle and that this would be a journey of love and cleansing and that we shouldn't worry. He would guide us every step of the way.

Next he began inviting each of us to come forward and kneel in front of the makeshift altar. He poured some of the thick brown liquid into a cup and held it out for each participant to drink. I was the second person to receive the ayahuasca. I went forward, knelt, and waited for the cup of thick brown liquid. Jorge extended the cup to me and I drank it quickly. It tasted like crushed prunes with a slight peppery flavor, but it was definitely sweet. I was surprised at this, having heard that ayahuasca was wretched in flavor and definitely not sweet.

I took my place against the wall and waited while the remainder of the people took their dose. Judith asked for a slightly smaller dose, and Jorge obliged her. After we all had taken our drink, Mike lowered the lights even more until the room was nearly pitch black. I could barely see in front of me.

I wasn't feeling anything in particular, yet, when I heard what seemed to be a bird's whistling outside of the house. I then noticed that my perception had shifted and that Jorge was making the bird whistles himself. But that didn't seem right either, for there was a definite echo of birds and a rustle of feathers apparently coming from the fireplace. I was beginning to wonder just what was happening. I reasoned that Mike and his assistant, Gladys, were also whistling and making the additional bird sounds. But yet that still didn't explain the feeling that there was a flock of birds around us.

While I was pondering this, the effects of the ayahuasca began to hit me like a ton of bricks. At first I became slightly dizzy. I noticed that each time I moved my head from side to side, the dizziness seemed to increase. I thought it best to sit as still as I could. So I did. Then I noticed a movement in front of me. Jorge appeared to rise from his lotuslike position on the floor and began to walk about the room, stopping briefly in front of each one of us.

When he first approached me, I was just feeling the ayahuasca, and as soon as I noticed him, I seemed to snap to vivid attention, very alert to what he was doing. I could barely see him, but he seemed to take on a glow. When I looked at him again, he appeared to me to be a very old Indian with long white hair.

The bird whistling and other bird sounds were still occurring and seemed to be coming from everywhere in the room. It dawned on me that the other participants were joining in. Then Jorge began to sing a lovely and simple song—the first *icaro*—a song of the jungle. As he did so he beckoned to me to give him something—something that was my pain or sorrow. I did. I offered something from my heart, and it appeared to me that a black cloud was passed from me to him.

Then he looked straight at me, or so it seemed, for the darkness had grown even blacker. He asked me, without saying a word, if I understood what he was doing. I heard him in my mind. Then he showed me a ball that appeared to be glowing in the palm of his hand. He asked me if I could see the ball. I said I could. I did. But it was impossible to see the ball, because the room had grown nearly completely dark.

Not only did I apparently see something that wasn't there, I heard a voice that wasn't spoken. It would have done me no good to hear him speak to me in Spanish, anyway. I somehow understood what he was "showing" me and what he was "saying."

The ceremony continued, with Jorge passing before each person and apparently asking each the same as he had asked me. At least that is how it appeared to me. Then Jorge asked us to close our eyes. I did. I then saw a variety of colors and spinning visions.

Now let me explain this a bit further. When I closed my eyes I didn't just immediately see visions. I saw the usual darkness you see when you close your eyes. But then there would be a crack of sudden vision, as if a lightning stroke had zapped across my eyes. Through the outline of the stroke, visions would appear. And nearly just as quickly, they would vanish into the darkness again. And then the process would repeat itself.

All sorts of images came to me. Most of them were from my childhood. I saw comic book heroes, Captain Marvel, Superman, Batman, and a wide range of television and movie heroes. I thought to myself, "I am on a hero's journey," and then the images would run through me. I was hoping that I would suddenly emerge in a different locale and that indeed I would be able to travel with my mind to the Peruvian jungle, but that didn't happen.

Then the first wave hit me. I didn't know what it was. I noticed a sudden burning in my stomach. Jorge had made sure that each of us was given a bowl in case any of us had to vomit during the ceremony. I definitely had not felt as if I was going to throw up. But to my surprise I suddenly felt the retch coming on and I threw up in the bowl.

It was horrible. But subsequently I felt at ease and more images ran through me. I then had the first sustained vision under ayahuasca. I saw myself as an eagle, and I suddenly had insights into the truth of the way of the planet. I saw as an eagle sees. Although it is impossible to describe my vision completely, allow me to tell you about it to the best of my imaginative ability:

The eagle soared. She flew high in the lazy hot afternoon sky. Her natural tendency was to look down as she flew, although with her omni-eyes she could see in every direction at once. Omnisphere was her visual world. She saw everything below as the order of natural motion. Wisps of clouds forming and vanishing appeared to her eyes as the natural rhythmic vibration of the great mother earth below. She swam through the cloudy wisps and felt the resonant harmony of the creation and evaporation of floating water drops.

Fields of grass undulated below her. The afternoon wind was stirring the body hair of the great mother earth. Even the streams, like rivulets of sweat on a great body, would echo the sunlight to the eagle's eyes. The streams moved in rhythmic swells as they flowed ever downward seeking the valleys between the breasts of the great mother. All was in harmony. All was as it should be.

The heat rose from the mother's skin in waves. The eagle felt them and then soared upward, motionless. Slight cold winds hinting at the coming dusk stirred the fine feathers on her underbelly.

There! There it was. A tiny disturbance. A perturbation. A movement below of something out of harmony with the mother's undulating surface. It was out of phase with all of the patterns below. There! There it was again. The eagle's attention was now fully focused. The omnisight was shielding over and the vision was sharply brought into focus on the disharmony. It was a small animal, no longer than a few inches. A mouse, perhaps . . .

The mouse darted into the sun for a brief moment. It was frantic and afraid. It was exposing itself for reasons that to it were not even clear. Its instincts were to remain hidden. What brought it out into the sun?

The eagle knew. It was the eagle's gift. A sacrifice from the mother. The eagle, her belly empty, her children crying in the nest a long way from here, swooped down in a dive that a jet pilot would envy, and hooked the mouse in her claws.

It all happened in a split second. The mouse hardly suffered any loss or pain. The eagle lifted from the ground and soared, opened her shielded omni-eyes, and once again searched for her nest and her children.

The eagle's quest was simple. Search for that which is out of phase, out of rhythm. It is the gift of the great mother earth. That was her food. That will sustain her and all of her children from now until the end of time. That is the truth which she will always seek. It is the truth of her own existence.

My wretched and gifted state lasted many hours. I had vomited around seven times and finally was feeling much better. The singing and whistling had continued throughout the ceremony. But then a new phase began. Jorge invited each of us to come forward and lay down at his feet. The maestro was about to heal each of us. I was again the second person to receive this healing. He placed his hands on my body and sang and danced around me. It seemed to last for a very long time. He poured some aromatic water over me, and I felt chilled at first but then warmed by the liquid.

Next I was asked to kneel in front of Jorge. As I did, I noticed, in the dim light that had now grown brighter, that Jorge was holding a baby in his arms. He appeared to me as mother with her child. He was cooing to the baby and giving it much love and attention. I wondered where the baby came from. But then the image of the baby dissolved and I saw that the baby was nothing more than my two books—the same books that I had given Jorge to bless. He told me that these books were my children and that they were born from my brain. They had to go out into the world and grow. I was very moved by Jorge's words. Then I sat up and Jorge embraced me.

As I sat there, Mike asked me if I had any questions. I told him about my visions and how I had felt. I told him that I understood much from the touch of Jorge's hands. Then Jorge told me that he was blessing my books and that I was very important to him. He told me that he would invoke me in the jungle when he returned to Peru and that I would feel something wherever I was. He was fulfilling my wish to time-travel, and he would attempt to bring my spirit to the jungle. I was invited to come to the jungle to participate in more ceremonies that summer.

After that I returned to my place against the wall, and each of the others came forward for his or her individual healing. By now the effects of the substance had worn off, and I was feeling much, much calmer.

Soon all of the participants had experienced individual healings. The formal ceremony was now over. Several people were up and milling around.

I walked over to Mike and Jorge.

"So you more or less understood the process?" Mike asked me.

I did. I told Mike that I realized that Jorge had become my maestro. He had taught me what the vibration was. I could feel his vibration so I now knew what the vibration was. It was physical. It seemed impossible for me to detect before I took the ayahuasca. But during the ceremony I had felt it.

And then I told him about how I began getting all of these visions. My visions were being bombarded by everything I have seen, television images, noise on the circuit, needless junk. It was like watching my neurons discharge garbage. Every negative thought I could think of came up. And then I saw my enemies and I realized that all my enemies are only the friends I have betrayed. So my enemies I respect, because they were my friends.

Jorge answered, "This is a tangible example of how our unconsciousness is guarded and makes us suffer."

"Now what I want to learn is how to deal with what I have seen so that it doesn't become my armor," I said.

Jorge's answer was surprising to me because I had learned it before. "The best way to get out of our unconscious desires is to become conscious of them, to become aware of them. I can help you in other sessions. I can guide you. Did you notice the desire to teach in the session; did you notice the educational aspect of the session? First you teach how to whistle and then you teach how to sing."

"I sang in tune, usually I don't sing in tune," I said.

Mike asked, "I wonder what the neighbors think of all of this?"

"I noticed that Jorge worked on my stomach. He seemed to work on me for a long time. Was there anything more I need to consider?"

"Just cleansing. I provoked the vomiting, too," Jorge said.

"I had both fear and joy in this experience. I had fear because of the retching and the dark side I had to go through to reach the light. And joy because of the revelation that came from it," I said.

"The more you get into it the less you have to vomit because you are cleansing out. Jorge didn't vomit at all," Mike said.

I explained that my dizziness and vomiting were due to a combination of things. I said, "After a while, I began to break out in a profuse sweat. I never felt that I was going to pass out. I am very sensitive to such things and I could have passed out, but I didn't. Then suddenly I would feel a burning in my stomach and I would feel a wave come over me, and then I would feel the drug even more and would throw up. I must have felt seven waves.

"Then the visions would come. Many times the eagle came. I was looking down, I was looking down as if from a high place like an eagle's flight. I saw the whole planet in a flash, and suddenly I realized that the earth was a gift, a gift from God. I realized that the eagle only hunts when it wants to eat or feed its young. I saw what the eagle sees when it hunts. All life moves in harmony with the motions of the earth, the wind, the clouds, and the breeze over the grasslands. When the animals below move in harmony with the earth, there is no danger for them from the eagle. But when an animal moves disjointedly, out of harmony with the rest of nature, the eagle sees it and immediately pounces on it.

"That is the gift to the eagle. The animal who moves out of harmony with nature is the sacrifice to the eagle. I realized that we today are moving out of harmony with the earth, so we shall be the sacrifice to the eagle.

"I always believed that my totem animal was the wolf. Now I realize that the eagle is also my totem animal. I didn't know this until now."

Jorge said, "You can find out many things. Did you do sessions with all the different shamans that you have met? It would be good."

I told them that I had done healing sessions with two other shamans, but those weren't with a group. Jorge asked me if they sang also. I told him they didn't. I explained that this was my first experience with the vision vine and that this session felt more as I had understood a shamanic session would feel.

"This is how we work in the jungle," Jorge said. "All negative forces have an exit, and they look to get out. The vomiting, the sweat, the diarrhea. If you get a little bit of diarrhea it is better to even help it along. It is better that for the next two days you stay on a soft diet, no animal fat, sugar, or citrus juices."

I mentioned the taste of the ayahuasca seemed to have prune juice in it. Did it? It didn't.

About two weeks after my first experience with Jorge, I participated in a second ayahuasca ceremony in the United States. I wanted to verify my first experience. The ceremony began with Jorge explaining that this was his last ceremony before he returned to his home. He told us that, in Peru, there are three types of medicine people: brujos, shamans, and curanderos.

Brujos are witches and they deal strictly with black magic. They cast spells and use medicine herbs to cause harm to others. Curanderos are healers, only. They do not use black magic and will not cast spells to harm any individual or any living animal or plant. They use what we might call white magic. In the middle world exist the shamans. They deal with both forms of magic. They will heal people with sacred herbs and they will also attack brujos with shielding forms of magic that redirect the black magic back to the source.

Jorge told us that all plants must be infused with the intent of the shaman before they will be effectual. Jorge infuses his herbs with his breath, which he says is his life force. He inhales deeply on a cigarette and blows the smoke making the sound of the wind as he breathes the smoke into the mixtures he prepares. He uses tobacco quite often, and I was surprised to see that his teeth were completely white, not stained with nicotine as I would have expected for a heavy smoker.

The ceremony began as usual with Jorge and Mike lowering the light in the room. He then called for each of us to come forward and imbibe ayahuasca. This time I was one of the last to drink. The mixture tasted quite bitter this time and appeared to be a little less viscous. I drank it all down, returned to my place, and waited while the ceremony continued.

This time I was disappointed. Nothing happened to me, and as I noticed by watching the others in the room, very little happened to anyone else. After around six hours of participation with Jorge and Mike whistling and singing the curing songs, the period of personal healing took place. It was then I listened carefully to hear if anyone else felt as I did. No one retched during the ceremony and everyone agreed that nothing was felt from the purge. However, I felt one of my chronic sinus headaches coming on and nursed it all during the ceremony. I asked Jorge to heal it for me when my turn for personal healing came up. However, I didn't feel or see anything that indicated that he had healed me.

Later in the week I had an opportunity to visit with Jorge before he returned to Peru. He again told me that I would be very welcome to come to his home in Tarapoto and then to Iquitos to visit and participate with him and his maestro, Don Solon. When I realized that he was quite sincere, I suddenly said, "Yes, I'll come." After I had left the farewell party for Jorge, I wondered to myself why I had agreed to go. But then it was April; I wasn't planning to go until June.

And now it was the end of June and I was back in the theater with Sandra and Daniel watching the movie. The movie shaman completed his dance-like movement around Nexy. Clearly, her consciousness had changed by the expression on her face. Somehow she saw herself in a different light. She had passed through her first initiation.

I then remembered my ordeal with the ayahuasca and how I had felt so ill. And yet I had felt that my ordeal was necessary. But why was it necessary and what did this ordeal have to do with initiation? Later that evening after the movie, I recalled my hypotheses. In a typical ceremony, shamans will use any device to alter a patient's belief about reality. Any device? I thought. Had Jorge used the ayahuasca to alter my belief about reality? Perhaps the ordeal I felt was of my own consciousness. Perhaps I really did not need to feel the experience as an ordeal. Perhaps it wasn't an ordeal after all, but just a realization of my own fear.

My mind raced on. Jorge, in taking me through a typical Peruvian shamanic experience, had actually initiated me. It dawned on me that shamans must prepare all patients by initiating them into shamanhood whether they wanted to be initiated or not! And now I realized what shamanic initiation does. It prepares the patient by altering the patient's belief structure, and that experience, especially if healing is required, always appears to be an ordeal because the patient must confront fear.

My first session with Jorge was no exception. He made me confront my own fears—in this case it was the fear of losing my mind. He made me realize that it was only my fear that held me from experiencing a full and healthy life. I, too, needed healing.

I had gotten into a rut. Even though I wrote about the magical world of quantum physics, I still remained, as I said in Chapter 1, a mechanical thinker. But this wasn't the first time that I had managed to pass through my own fears. I then remembered the very first shamanic ordeal I had passed through a few years ago. I didn't know that my ordeal had anything to do with shamanism at that time. It is only now that I realize that this experience also made me confront my own fears and had altered my own belief structure. But what was most surprising about it, even though it was an ordeal, was that I had felt no pain although everything happening to me said there must be pain and injury.

I now realize that this experience was a typical technique used by shamans to encourage their patients to undertake journeys of such physical demands that they would at first sight seem impossible. My journey was a short one, only twenty feet or so. Nevertheless, it demanded that I go beyond any preconceived boundaries that I may have constructed for myself.

It was the spring of 1984 and I was living in San Diego, California. My mind drifted back.

A FAMILY FIREWALK

In that year, a new rage had hit the California coastline. Crowds of people were attending "firewalking" seminars, led by Tony Robbins, in which the participants were actually encouraged, after listening and paying for a seminar on positive thinking, to walk across burning coals of fire. By the time I had found out about it, several hundreds of persons, old and young, had passed their bare feet across the length of a stretch of hot coals around six feet wide and about twenty feet long.

I must admit that even though it was impressive to me that so many people in the United States had undergone such a feat, I had no desire to firewalk myself. Fear, reluctance to pursue a less-than-scientific venture, boredom, or whatever else kept me from inquiring into it. I rationalized to myself that, somehow, if all those people were doing it, it must have been a trick. My wife, Judith, however, was fascinated and said to me that it was my duty as a scientist to check it out.

Later that week my son-in-law, Scot, who is a professional photographer, told me, it just so happened, that he had talked with Tony and would be taking pictures at the next firewalking ceremony in San Diego, and that he could get several passes for Judith, my daughter Leslie, who was married to Scot, and me to attend.

I was admittedly stuck. I had no excuse not to go and so, with some reluctance, Judith, Leslie, my stepdaughter Shawn, her boyfriend Jamie, and I attended the firewalk seminar. We were told immediately that, just because we were there, we didn't really have to walk across the coals. We wouldn't be forced to do so, and it was dangerous to do so unprepared. The real purpose of the seminar, Tony told us, was to encourage us so that we could do practically anything, even if it seemed impossible.

After several hours of listening and participating with about one hundred others in a large room at the Convention Center in downtown San Diego, a group of very positive people—all of the participants—were led from the second floor of the center down to a roaring bonfire in the middle of the outdoor ground floor. I wondered, as we walked down the stairs, each of us chanting, "yes, yes, yes," etc., just how the seminar organizational committee got permission to put on such a display in the very middle of the prestigious San Diego Convention Center?

We were led into a circle, and each of us had the chance to walk around the fire, which at that time had flames leaping over eight feet in the air. The heat was immense, and I'm sure that each of us began to have considerable second thoughts.

Then, as we gathered into a circle around the fire, the coals were raked over and spread out in a long strip, wide enough to walk across. The flames were damped out by the spreading of the coals, but the heat from coals in that little strip was quite apparent.

Before any of us had any time to think about it, and with the encouragement of the seminar leader's assistants, we found ourselves hypnotically forming into a line and beginning a slow march across the coals. And, in a few minutes, I and all of my family together, with nearly all of the remaining participants, found ourselves walking across the coals.

As I approached my turn to walk across the coals, Tony, the seminar leader, whispered in my ear, "Remember, look up as you walk and not down at the coals and keep reciting to yourself as you walk, 'cool moss.' " I then started my walk reciting to myself "cool moss" as I took each step. I didn't hurry as I walked. In fact Tony warned us not to run or move too quickly across the coals. I distinctly remember how each step felt. Not once did I feel any heat coming to my feet, even though I felt heat surrounding me and reaching my face.

I did end up with a few blisters, however. But no burning was felt at all as I marched across the coals. How did we all do it?

My first thoughts were that the secret was based to some extent on some basic laws of physics governing heat flow and heat capacity. These same laws govern the reentry of the space shuttle into our atmosphere from deep space. Without the discoveries of the new technological advances that make reentry from deep space possible, I doubt that any explanation of firewalking would have been forthcoming. I believe this since no physicist I know of has[3] heretofore advanced any plausible explanation. Let me offer one.

To grasp the secret, at least as far as the basic laws of physics go, one must look at just what we do understand about heat and temperature. Before I tell you this, let me remind you of an experiment with electricity you probably saw when you went to school.

Remember when that zany professor of physics asked a young college or high school student to come to the front of the room and stand on a wooden box while he attached an electrode coming from a big machine called a Van de Graaff generator to the person's unsuspecting arm? Or perhaps he asked that person to take hold of a wire or something like that. As soon as the generator was rotating with full power, the students were all amazed to see the hair on the volunteer's arm or head raise up as the hair became charged with electricity.

The professor then told you that around fifty thousand volts of electricity were flowing through the volunteer's body. Did you wonder, if that much electricity was flowing through his or her body, then why didn't the person just burn up and die like an electrocuted prisoner?

Well, it turns out that more than high voltage is needed to really electrocute a person. One must also have current or electrical charge moving through the person's body. Voltage is just a charge pusher. Without any charges or with just a few charges, even a high voltage won't accomplish very much.

But even if a large amount of charge is somehow available, in order for that charge to move it has to have somewhere to go. In other words, it must reach a storage tank—a place where charge can accumulate. The measure of a body's ability to take on electrical charge is called its electrical capacity. The amount of charge that can be stored in or on a body is the product of that capacity and the voltage. So if the voltage is very high, but the capacity is very low, not much charge will pass into the body.

[3]It is possible that a physical explanation has appeared or that one will appear before this book is published. If so, I apologize to any physicists or physiologists that I may have slighted.

Even if the body does not have much electrical capacity, charge could still pass through it. However, every ordinary substance offers some electrical resistance to the movement of charge. Without resistance, any tiny voltage acting on a charge would be able to push it a great distance. Superconductors have no electrical resistance, but ordinary materials always resist electricity. So with a great resistance, even a large voltage will not be able to push much current.

Our bodies have large resistances to the movement of electrical charge and small capacities for accepting electrical charge, so that even with fifty thousand volts, not much current passes into a person's body when an experiment like the one described above is performed.

In a similar way, our bodies are resistive to heat and won't accept much heat passing through them under certain conditions. Today we can build ceramic materials with very low heat capacities and high heat resistances. Now if one takes a space-age ceramic tile from a reentry space vehicle and examines it closely, one finds that it indeed has a very small heat capacity and is able to withstand a very high temperature. As the space vehicle enters into the atmosphere and its wings reach very high temperatures, the low capacities of the protecting tiles enable the inside of the ship to remain cool. Not much heat passes into the interior of the ship because the tiles have very low heat capacity and high resistance to heat. One engineer demonstrated this ability by taking a tile out of an extremely hot oven and holding it in his hand with no damage.

The relationship between heat and temperature is similar to the relationship between charge and voltage. The heat flowing into a body is the product of the heat capacity of the body and the temperature of the body. Just as a high voltage doesn't necessarily produce a high current, a high temperature doesn't necessarily produce a great amount of heat passing from it if the body has a low heat capacity. Also, even if there is a great amount of heat coming from the fire, it is not necessarily the case that the heat will pass into the skin if the skin resistance is high. And if those coals also have high temperatures but small heat capacities, not much heat will be present in each coal.

The overall amount of heat coming from all the coals will be quite high, but the amount of heat coming from those fewer coals, covered by the feet as they pass across the fire, will be sufficiently small to not do much damage. The coals are to the feet as the reentry tiles are to the underskin of the space vehicle.

Now that I have told you that, are you convinced? Well, I'm not, even though I have "explained" it. After all, I actually walked across those coals, and I want to tell you that I didn't feel any heat passing through my feet or even any heat on the soles of my feet. The walk felt as if I were trodding across a field of crunchy cool moss.

I now realize that my scientific background and the explanations I have offered do not work. My ordeal with fire was a shamanic initiation. In crossing those coals I had confronted my own fears that it was impossible to enter fire and not be burned. I had actually been through a consciousness-altering experience, one that made me realize that I was capable of doing more than I had realized.

As I researched the practice of firewalking, I found out that firewalking was a technique used by shamans to enable them to enter into trance states. It is a common practice in many parts of the world. In Japan, Buddhist monks will walk across coals on festival days. Afterward, ordinary people follow in their footsteps. The monks recite special prayers to themselves as they walk. I can imagine that they are incantations similar to our "cool moss" chants.

My second hypothesis appears to enter here. These monks were seemingly violating the laws of physics. Were they simply using the observer effect and seeing the world in a different light? Hypothesis 3 states that they were perceiving reality in an altered state of consciousness, which allowed them to violate the laws of physics. Perhaps those chants were mind altering in their effect on the body. Perhaps the firewalkers were violating only the classical laws and not, at some deeper level, quantum physical laws. Or perhaps there was a device used to make everyone witnessing the events of the firewalk believe that the fire was hotter than it actually was, thus confirming hypothesis 4.

Although I have explained the phenomenon in terms of physical principles, let me now examine some physiological facts. Some medical physiologists have written[4] that the whole explanation lies in the toughness of the soles of the feet of the firewalkers and the extremely high tolerance for pain that these people have. I can assure you that my feet are quite tender and my pain threshold is quite normal. In fact, let me remind you that I didn't feel any heat coming into my feet, although I felt quite a bit of it on my face and arms coming from the surrounding coals along my walk.

Some have suggested that the firewalkers put some kind of substance on their soles that protects them from the heat. But no one knows of the existence of any such substance, and for that matter, none of the people I

[4]See Weil, Andrew. *Health and Healing. Understanding Conventional and Alternative Medicine.* Boston: Houghton Mifflin Co., 1983.

saw firewalk, including yours truly, applied any kind of ointment or other substance to our feet.

Some scientists suggest another mechanism, called the Leidendorf phenomenon, that protects the soles of the feet with a microlayer of gas formed by the evaporation of sweat from the soles. Let me explain. If you put a small drop of water on a very hot skillet, you will notice that the drop will begin to skip and dance across the skillet, seemingly maintaining its droplike form for what seems like a very long time. After a little while the drop will vanish in a puff of steam.

What happens is that a microlayer of steam forms at the interface between the skillet and the drop. This microlayer, like a ceramic tile, has little heat capacity and high resistance; consequently little heat passes into the center of the drop. However, after a while, with the microlayer evaporating off into the air, the drop shrinks until a point where the surface-to-volume ratio is insufficient to protect it any further. In other words, there is more surface for the heat to cross into the drop than volume present, so then it goes poof.

In a similar manner one believes that the sweat-gas microlayer protects the soles of the firewalker's feet. However, many firewalkers walk through the coals ankle-deep (although I didn't do that), with the coals passing well over the sensitive skin on the tops of the feet. Indeed they seem to shuffle through the coals as if they were shuffling through a field of wet tall grass.

I don't believe the microlayer theory for another reason. In 1974, a physicist from the Max Planck Institute for Plasma Physics in Munich made careful measurements of the temperatures of rocks in a Fijian fire pit and of the feet of the firewalkers who crossed those rocks. Now rocks, as you know, heat up very slowly, but when they get hot, they're hot! Rocks have high heat resistances and large heat capacities. Consequently, while they heat slowly they do retain heat. That is why rocks are used in some heating vents in homes in New Mexico during the winter.

The physicist measured the temperatures of the hot rocks and found that they were 600 degrees Fahrenheit; however, the feet of the firewalkers never rose above 150 degrees. When the physicist placed a piece of callused skin from one of the firewalkers' feet on the rocks, it carbonized almost immediately. He concluded that no explanation from the laws of physics would suffice.[5]

So if there is indeed a change in the resistance and capacity of the skin

[5]This was reported in Doherty, J., "Hot Feat: Firewalkers of the World." *Science Digest*. August 1982, pp. 67–71. Also see the letter to the editor in response to this article, "Atheist Firewalkers." *Science Digest*. November 1982, p. 11.

of the firewalker's feet, it must be caused by some ability of the human mind. I certainly believed and felt that I could walk across the coals unharmed and perhaps that belief altered my consciousness and enabled my mind to influence the soles of my feet.[6]

Changes in the functions of local nerves might suppress the activity of neurochemicals that mediate pain and inflammatory reaction to heat. Experiments in hypnosis show that suggestion can cause the skin of a subject to show heat damage when no heat has been applied and can cause the skin to show no heat damage when heat is applied.

Still, after the experience, I wasn't sure how I could ever really bring together my scientific understanding and my actual experience. It was as if the two had taken place in different worlds. Perhaps the experience enabled me to cross over from the scientific world of thoughts and sensations into the shamanic world of feelings and intuitions. I later found out that I wasn't that far from the truth.

A PHYSICIST'S SHAMANIC INITIATION

In one of my worlds, I was here in Lima watching a movie. Although I was seeing the young heroine, Nexy, undergoing her rituals, my attention was on Miguel. I kept feeling that Miguel's life and mine were somehow intertwined. He was just an actor following a script, and I had begun to feel that I, too, was following some kind of script. These flashbacks were telling me something about destiny. And at the same time I was undergoing my own ritual. There was a connection between initiation and destiny.

My mind kept flashing through time. Just a few days before, I had been in Cuzco to participate in the famous winter solstice ceremony called *Inti Rahmi*. It was a special time.

The night before the ceremony, I got into bed and began thinking about healing again. How does a shaman prepare a patient? Perhaps there are stages of preparation. Certainly when a physicist prepares an experiment, he must pass through several stages before he sees any result. As I tried to

[6]I recently learned that a physics professor at the California Institute of Technology and his students walked across burning coals in a demonstration held at the Institute. They certainly had no mystical or spiritual belief going for them. They were convinced that the laws of physics were perfectly adequate in explaining why they would suffer no damage to themselves. I would suggest that it doesn't matter what you believe is the reason for a successful firewalk, physics or metaphysics, you need to have faith in something to accomplish the task. Perhaps just the belief that you can accomplish the task is necessary. It may not be a sufficient requirement for a successful walk across the coals. But without it, I am sure one would get badly burned.

fall asleep my mind kept turning over and over the thought that a shaman must prepare the patient in ways that we have never been exposed to in Western medicine. But how? What was the preparation? The word "truth" kept running through my mind. What was the truth?

My mind was racing. I remembered my early childhood experiences with magic. Did I really slide down the stairs? I remembered the fly in Nepal. Did I really become one consciousness with it? I remembered the firewalk. My logical mind could not really figure out how I managed to walk across any of these events.

But I was prepared for them. My whole life to that very minute, lying in bed with the Inti Rahmi ceremony beginning in the morning, was beginning to appear to me as nothing but a preparation—an initiation that I was destined to follow. For what purpose? I had taken the steps to leave a comfortable job, security, and all the benefits that go with a tenured academic position. And yet I still wasn't sure why I did it.

And now in the movie, Nexy was going through her initiation. The shaman was dancing around her, and he had his arms widespread. He looked like a gigantic bird, an eagle, as he danced around Nexy. And then another memory flashed in my mind. It was my first initiation experience with the North American Indian shamans and the first time that I was to become acquainted with one of my totem animals, the eagle. (You'll recall my brief discussion of totem animals in Chapter 3.) I was living in Santa Fe, New Mexico.

THE EAGLE'S INITIATION

It was midwinter, February 1989, just a few months ago, two months before my first vision vine experience. It seemed an eternity. The clear cold night of Santa Fe, New Mexico, showered me with floodlights of stars.

I entered a tepee in back of a friend's home and sat down on the chilly ground. My feet hung over the edge of the fireless fire pit. Ed McGaa, visiting the medicine woman, Jamie Sams, had brought me there. Ed was an Oglala Sioux Indian, author and former Viet Nam Marine combat pilot. He began his chant as he stood in the center of the fire pit that remained cold and dark. He held the hollowed eagle-bone whistle between his teeth and whistled and sang a chant that I had heard before and yet had never heard at all.

I heard the sound of a drum in the background. But no drummer was present. I heard the rustle of large wings. But no bird was present. There was no one else in the tepee except the shaman and me.

The constant rhythmic beat of the drum and the wings, as I sat with my arms around myself in protection from the cold, brought me to a state of timelessness. Where was I? When was I? Time and space seemed to dissolve around me, and I was left with only the sounds that I heard and the sounds that I did not hear.

He placed his hands on my head, and at that moment I felt the claws of an eagle clutching my scalp. I felt them pulling me, lifting me, carrying me upward, flying over, into the night void splashed with stars.

It was done. We left the tepee and returned to the house. As we walked through the starlit field, the Indian, who was in fact young but, in another reality, very old, asked me, "What was your experience?" I said that I had the distinct feeling that an eagle had entered the tepee and had carried me away. The shaman said, "I know."

As I watched the movie shaman I wondered, how could Ed know what I felt and what I saw? The movie shaman was still circling about Nexy. Did Nexy feel what I felt? And then another initiatory experience flashed in my brain. I saw the eagle again. I was going through a range of emotions, laughing and then crying and then sudden clarity.

THE SWEAT OF THE QUESTING EAGLE

Earlier, when I had met with Ed McGaa, I had asked him, after he had initiated me in the tepee, "Did I check out with your ancestors?" He told me that I did, and he said he knew my Indian name and that I would be given that name in a ceremony later. I wondered when that would happen.

It happened sooner than I expected. My friend Jamie Sams had invited Judith and me to take part in a traditional Seneca purification ceremony. Jamie belongs to the Wolf Clan of the Seneca tribe and is a versatile and talented author. She has also been initiated as a seer of the Kiowa Buffalo Society and is a Choctaw-trained Pipe Carrier, authorized to perform healing rites using the sacred pipe during the purification ceremony.

Jamie leads these sessions fairly often, but in order to take part in one, it is necessary to be invited. We both felt quite honored at the invitation. When Judith and I arrived at the ceremony grounds just south of Santa Fe, New Mexico, we didn't really know what to expect. Jamie hadn't arrived yet, so we walked out to the lodge grounds. There we found two people tending a fire just in front of the lodge. Jamie had asked us to bring five pieces of firewood to keep the fire going. We did.

The day was warm as the afternoon sun lowered in the sky. Some clouds had been blowing in from the mountains, and soon the sky was nearly completely overcast. New Mexico's sky is something to be appreciated. It has a magical quality, a feeling that one has entered a primal state of consciousness, one that existed even before human beings walked the planet.

I looked at the sweat lodge carefully. It covered a circular piece of ground about fifteen feet in diameter. Inside the lodge a pit had been dug. White sheets were placed on the ground surrounding the pit for us to sit on. To construct the lodge, twenty or so thin willow trees, perhaps about ten feet in length, each tree an inch or so in diameter, had been first cleaned of branches and leaves. This made the trees into quite bendable wooden poles. Next, the poles had been placed into equally spaced dug out holes forming the circumference of the great circle.

Then the poles had been bent over toward the center of the circle. In this way a framework had been constructed in the shape of an inverted bowl. Next, old blankets, thin mattresses, opened sleeping bags, and whatever of the like had been placed over the frame. The lodge's construction was nearly lighttight and airtight. The height of the lodge at its center was no more than three and a half feet, so one could not stand inside. A small doorway was provided for the participants to crawl inside. I had guessed that when everyone had taken his or her place, the doorway would be closed with a blanket.

I wondered how we were to all stand the heat during the ceremony. I had expected that pieces of burning firewood would be placed in the pit inside, and so I supposed that soon we would be smoked out.

Soon, Jamie arrived. She told us that the ceremony, which was to begin at 4:30 that afternoon, would begin as soon as seven other people arrived. By 5:15 the other seven had not arrived. (I found out later that they all had gotten lost finding the site.) So we began.

Jamie explained to us that the Purification Lodge is round and represents the womb of the Earth Mother. We are forced to enter the lodge on our knees in order to humble ourselves and acknowledge that we are no greater than any other life-form in the planetary family. When we finish the four endurances, each consisting of songs, prayers, and purifications, we leave the womb of our Planetary Mother in order to be reborn into our new state of being.

There were four of us in the lodge and one person outside watching the fire. We had removed our outer clothing before entering. I now wore a pair of swimming trunks and sat in the East. Judith had on a tee shirt and bathing suit and sat in the North. The third person with us, a Filipino woman named Terry, wore a bathing suit under a towel wrapped around her. She sat in the South. Jamie entered the lodge last and was sitting at the Western door. She wore a loose-fitting white cotton shift.

The ceremony began with the smoking of the sacred pipe. Each of us puffed on the pipe and handed it around the circle to the next person. The smoke tasted very sweet. As we inhaled or puffed on the pipe, Jamie chanted song-prayers to the various powers of the seven directions: East, South, West, North, Above, Below, and Within. I was still wondering about the flaming firewood and whether we were going to be consumed by smoke.

Jamie assuaged my concerns by calling to her assistant outside for the first seven stones. That was it. Hot stones from the fire were placed in the pit, not the wood itself. Soon the stones were put into the pit and the doorway was shut. I noticed the nearly complete darkness and then felt the heat from the glowing stones. I had never really seen stones glow before and wondered just how hot they would get.

I didn't have to wonder long. Soon I had broken out in a sweat that was not to let up. Jamie threw some rosemary oil on the stones, and the scent soon filled the space. The chanting began once again, with each of us following Jamie. Jamie then threw some water on the stones. The steam filled the lodge and the heat increased evermore.

Jamie then said that she had been given our medicine names by her Grandmother Ancestors and that each of us was to receive such a name in this ceremony. Terry was called Hummingbird Woman. After receiving her name, Terry was asked to identify with the name and see just how that spirit was her ally—in other words, to verify that the medicine name which represented her spirit was correct. She spoke English hesitatingly and consequently didn't really address the issue.

Jamie told me next that I was given the name "Questing Eagle." She asked me to identify with it.[7] I felt that the name was appropriate to me. I have always been seeking knowledge and attempting to find the highest truths available. I am also interested in the knowledge that can't be seen directly, hence my studies in physics. So I felt completely at ease with the name.

Judith was called Rising Moon Woman. She too felt identity with the name and said that for her it symbolized feminine beauty and the dream work that she does during her sleep-time.

Jamie's medicine name is Midnight Song. It, too, represented a power

[7]I should point out that receiving the name "Questing Eagle" occurred a few weeks after the time I was initiated by Ed McGaa in the tepee but before my first experience with the vision vine and my flying-as-an-eagle vision.

of the night and the mystical sense of the chant and was given to her on a vision quest twenty-five years ago. We all felt at ease with our names.

Soon the first ordeal of heat was ended. The doorway was opened and a welcomed breath of fresh air entered the lodge. None of us were to leave our places, however. Jamie called for three more stones, and they were placed in the pit. The doorway was closed again and the second endurance began.

Now the space felt even hotter, and we all were perspiring feverishly. Jamie asked each of us to make a prayer for the planet and our own place on it. My prayer was simple and surprising to me since I hadn't really thought about it before. The words just seemed to jump into my mind. I asked to be healed of all unnecessary pain and suffering so that I would be worthy and capable of bringing the light of knowledge to those that wished to witness what I had learned.

After Judith said her prayer, Jamie immediately focused attention on me. She said that she now realized why the other seven people had not come and that I needed to be healed. She did see just what I needed, even if I didn't know what it was at the time. I simply stared out blankly in the barely lit space of the lodge. She asked me to lie down for the rest of the ceremony as she questioned me about a blockage in my chest. Soon I felt a wave of sadness and grief overcome me. But the second round of prayers was now over and the doorway to the air was opened once again. I sat up briefly during the interlude.

After four or five more stones were brought in, the doorway was shut once again. I was asked to lie down and place my hand over my chest near my heart. Jamie said she felt what I needed and asked me to put my finger through the hole that was in my heart! I tried but couldn't find the right place. She then crawled past Terry's position and made her way to my side. She poked painfully into the ribs over my heart and said, "There, what does that pain in that spot bring up for you?"

I immediately broke into sobs as I thought of my son Michael, who had been killed by a drunken driver over two years ago. I said, "Michael." She said, "I know. Michael is here in this space with us now." I couldn't believe this in spite of the sobbing and sorrow I felt. But Jamie was persistent. She said, "He is telling you that he is all right and that he hasn't left you. That he is helping you, channeling information to you that will guide you in life as well as in sharing your wisdom through your books."

Jamie then asked me to get rid of my doubts and uncertainties about this. She psychically felt my disbelief and hesitation. She put her hand over my liver and said, "I feel some anger here over Michael's death. Shout out your anger." I didn't know what to say. She said it for me, "Damn it, why did you leave me?" I repeated the words several times, screaming them loudly, and felt my anger and sadness beginning to flow and subside. Then Jamie said, "Michael wants you to have more fun in life." I laughed and remembered that fun was something that Michael pursued at every turn. Then Jamie said, "Michael said remember the moped, dad."

I was dumbfounded. I had never even mentioned any incident about the motor scooter I used to have and the times I took Michael for rides on it with me. These were the days before the birth of my fourth child, when I took all my three children, Leslie, Michael, and Jackie, and my wife, Elaine, with me on laughing, exciting rides down the streets of La Jolla. Those were fun times and Michael always enjoyed them. How had Jamie known this? The answer was Michael told her. Shamans do enter the land of what we call the dead. They hear and see the spirits of those who have passed in the spirit world.

I was still feeling incredulous. Then Jamie said, "Michael said remember the three oranges." I remembered the classical music I used to play called "The Love for Three Oranges," by Prokofiev. I used to play that music in the living room. I didn't remember Michael listening to it.

Jamie elaborated when I told her I wasn't sure what the oranges meant. She said, "It has something to do with gravity and juggling three oranges. Michael said that this has something to do with the nature of forces in the universe."

Jamie then let up pressure over my heart and offered a healing prayer for me. She acknowledged the presence of the three women and myself and the healing power of the feminine energies. Then the third endurance ended as the doorway was opened once again for needed fresh air.

The fourth round of the ceremony was conducted in a lighter vein. We all sang sacred songs and gave prayers thanking Great Mystery for healing our hearts. When we finally emerged from the lodge, the sun was setting and the clouds that had previously covered the sky with gloom had nearly completely dissipated. The sky took on a variety of pastel pink and purple colors. We all felt that we had purified the burdens of our heart's pain with every prayer and song. The sky was a reminder of our purification and the healing that can occur when we return to the womb of the Earth Mother.

The pink of the New Mexico sky faded into the red sky of the movie jungle. I looked up at the screen. Meliton, the shaman, was finishing the ritual with Nexy and Miguel. They had been sitting as Meliton chanted. I thought of Jamie's chants and what she had taught me.

It was a valuable lesson. It wasn't an easy one for me to learn. Her awareness of the presence of my son, Michael, was impossible according to every law in physics I had thought I had understood. She had tuned into an archetypal spiritual world, probably one that existed inside of me. She was asking me to stretch my belief system. I had to do this to heal the pain I felt over Michael's death.

But my mechanical mind was still working. I still didn't quite believe what had happened to me. Perhaps it was real, and then again, perhaps not. Was the shaman really seeing into another arena of space-time, differing from the physicist's but nevertheless somehow connected? I recalled my sixth hypothesis (Shamans enter into parallel worlds) and my ninth hypothesis (Shamans enter the death world to alter their perceptions in this world). Or was the shaman just a trickster and, following my fourth hypothesis, using a device to alter my belief about reality. Maybe this had nothing to do with physics and I was a dupe.

I watched Meliton, the shaman in the fantasy world of the movie. Was Meliton a real shaman? Or was he just an actor playing the part? Was I missing an essential point of shamanhood? Reality and fantasy were not easy for me to separate. But was that the very point I was missing? Perhaps reality and fantasy are much closer than my physics training would allow me to see. Perhaps the whole thing is some kind of magic trick. I had always felt this when I studied quantum physics.

I remembered my meeting with the trickster shaman, several months ago in South Dakota, before I even knew I was coming to Peru.

THE TRICKSTER-SHAMAN FROM ROSEBUD

Ed McGaa had telephoned me. He invited me to come to the Pine Ridge Oglala Sioux Indian Reservation in South Dakota. He promised to find me a full-blooded Oglala Sioux shaman who would lead me in a ceremony. I accepted and flew to Rapid City the next day. Ed met me, and after a few days together we made our way by car to Pine Ridge.

When we arrived there, Ed introduced me to Paul Little, a half-blooded Sioux living on the reservation who is also a council representative. Ed told Paul what I was doing there on the reservation and asked Paul to "find me a shaman."

Ed had to return to his home in Minnesota, so he left me in the hands of Paul. Ed had made arrangements for me to stay in an apartment on the Pine Ridge Hospital grounds. These apartments were kept for visiting doctors and nurses. Paul accompanied me there, and as he drove off, he promised to find me a shaman on the reservation.

It was around seven o'clock in the morning on Saturday when I awoke. At 7:45 A.M., I decided to walk out through the compound. As I made my way up the road, I saw Paul Little driving up in his pickup. He had what appeared to be an old drunk sitting next to him. The old man's skin appeared to be brown and caked, looking as if he was just awakened from the top of a garbage dump. I had the impression that he smelled bad and his breath was stenched with alcohol. He also seemed quite dirty, looking like a bum from skid row. Paul said to me, "I want you to meet my friend here; Doug White, he's a medicine man."

At first I thought that Paul was kidding me. I literally thought that he must have dragged Doug from the nearest garbage pile. I wondered then what had I let myself into by asking to meet with an authentic Indian shaman? I, having a cleanliness fetish, could barely make myself shake his hand. Later I was to realize how incorrect my impressions of Doug White were.

Paul then asked, "Could we come into your place and talk." I was feeling very uneasy about this, but I decided to take a chance and I let them in.

As soon as we got in, I believed that Doug smelled as if he had been on a binge. He also appeared to have not bathed for quite some time. He told me that his wife had just kicked him out of the house. I couldn't guess why.

As we talked, I found out that Doug was a full-blooded Indian from the Rosebud Reservation in South Dakota. The Rosebud Reservation lies just south of the Pine Ridge Reservation, a few miles north of the Nebraska border. The Rosebuds like the Oglalas (which means "scatter-their-own") are Teton or Titonwan[8] Indians, which means that they are Plains dwellers. Rosebud signifies the location of the reservation but not necessarily a tribe.

Doug was sitting on the couch drinking coffee that I had made from instant and began chain-smoking cigarettes. He told me that he was going up to Canada, that Paul was going to take him there, and that he was going to do some "stuff" up in Canada. He sounded very unconvincing to me, and I continued to think that Paul had brought me a loony drunk rather than a full-fledged holy man. He didn't seem very holy to me at all.

Compared with my images of shamans, Doug seemed a long way off. Even though Doug appeared to me not to be what I had hoped for, I decided to ride this out. What the hell, I was down here and there was no point in just flying back to my home in New Mexico. I talked to him, attempting to find out just how he worked as a shaman or holy man. Most of the conversation was incoherent. He seemed to be bent on telling me about his wife and his stepsons and how they had beaten up on him. He

[8]See reference to *The Sioux* in bibliography for this spelling.

told me that he couldn't take that woman anymore. He repeated his plight to me, reminding me again and again that Paul was going to drive him to Canada.

I wondered why Paul was being so generous with his time and money to take Doug on a more than two-thousand-mile round trip. Paul later told me that Doug was powerful and that he had asked Doug to perform a special form of ceremony to help him. Later I found out that Paul had suffered a recent stroke.

When I had come to my wit's end dealing with Doug, suddenly he said to Paul, "Let's go." I was relieved to see them go but felt frustrated that so little had been gained by my trip to the reservation. But then Doug said that he was going to do a special *Yuwipi* thanksgiving ceremony for Paul that night. He said that he had to keep Paul well. Not knowing what that would mean, I said, "Oh that's nice," or something just as innocuous. I did not ask them if I would be welcome there. I was hoping that they would invite me. They didn't.

Just as they were about to drive off, I began to worry. Perhaps I wouldn't be invited to a ceremony after all. I put my hook in. I said, "Do you think that I could come to the ceremony tonight?" Now that is what I really wanted to do. Although I had been through a vision vine ceremony already,[9] I had never seen a Yuwipi ceremony.

Paul looked at Doug and asked him what he thought of that. I wondered if he even heard Paul, but then he said abruptly that it would be okay. So I was now invited to a ceremony that very few white people have ever seen.

I then asked Doug what kind of ceremony this would be. Would it be an actual Yuwipi ceremony? Normally during a Yuwipi, the shaman is completely bound by leather tongs and covered tightly with a buffalo skin. This skin is then tied and bound so that the shaman cannot escape from his cocoon. Ed had told me earlier that I had to go to a Yuwipi ceremony while I was on the reservation. Well, Doug wouldn't tell me exactly, but did sort of indicate that it would be a Yuwipi.

It turned out to be a modified Yuwipi ceremony. It was Yuwipi in every way but one. The people there didn't tie Doug up. They didn't do that.

So, I asked myself, was this apparently foul-smelling drunk, sitting in my apartment, a real shaman? I was relieved to see him go. As they were driving off, I wondered if I would ever see them again. I felt like a fool. Paul told me that they would be back for me at sunset.

[9]This took place just before my first sweat lodge ceremony with Jamie Sams and after my first vision vine experience.

THE SHAMAN WITH THE FLASHING LIGHTS

My earlier time with Ed McGaa had taught me something about the Sioux Indians. They always keep their word. Although my first encounter with Doug was less than I had expected, I knew that Paul would return. So about seven that evening, as dusk rapidly approached, I went outside of my apartment, and there Paul was, sitting in his pickup truck. Sitting next to him was a man with a natty brown shirt on. He was wearing a naval sea captain's cap. It even said "captain" on the brim. I laughed at that and I walked over to see who this was. It was Doug, the shaman. He was all cleaned up. He was wearing clean clothes and he had shaved. He also smelled high of Brut, the men's cologne.

I was amazed at his transformation. He then wanted to talk with one of the nurses in the hospital. I followed Doug and Paul into the emergency room and introduced him to a nurse named Mary that I had met earlier. Well, Mary recognized him. Seeing that the nurse he wanted to meet was not there, Doug asked for a telephone, and finding out where it was, he left me standing with Mary. Mary pulled me by the elbow to a corner of the room. She whispered to me, "You better watch out for him."

I said, "Why?"

"Because he's drunk."

I knew from his behavior that he wasn't drunk. But I then realized why Mary believed that he was. It was the Brut. He had smeared it all over his body, and it made him smell of alcohol.

I then realized that I may be dealing with a *Heyokah*, a trickster-shaman. I realized that he probably wasn't even drunk when I met him that morning, he just smelled high of Brut.

Doug returned and we walked outside. Doug approached me and began acting as if he were drunk. He asked me if I could help him out with some money. I looked at him, and, again, he appeared as if he were a bum. I realized that Doug was getting my emotions all upset. He was pushing all of my buttons. I didn't even realize that until several days later. Somehow Doug knew just what my buttons were. And money was definitely one of my buttons. He knew just what to say to upset me. But how could he know?

I then realized that Doug had quite a lot of shamanic power, more so because of his abilities to push my emotional buttons and to play the tragic clown role so well.

Soon we left the compound. I followed them in their pickup back through a windstorm in the plains. It was dusty and blowing a hot wind, as the sun was beginning to set. I could barely see their truck ahead of me, the dust storm was so bad. Yet, I managed to follow Paul and Doug in their pickup back along the main road threading through the reservation, toward the town of Oglala. We soon reached a dirt road leading off toward the left of the main road, leading to a house up toward the hills of the badlands of South Dakota. With each mile, I found myself growing more and more suspicious. As we left the main road, it was getting darker and darker, and at the same time more and more brush and plant life, shrubs and trees appeared to be looming around me. I was becoming frightened, and I felt as if I were taking a journey into the darkness of my soul.

We finally arrived at a small house nestled against the foothills. It was very dark now. There are no street lights anywhere on the reservation except near the hospital compound.

I walked into the house. There were a number of people assembled. Against one side of the living room nestled a small wood stove. There was no other heating available. The windows were totally covered with blankets and curtains. Folding chairs were arranged along three sides of the room, making a square with one side missing. I noticed that the walls were colored pink. I also noticed that every mirror in the house was covered. My fear began to grow.

Sitting were three Indian women. I then asked one of the women about the covered mirrors. She told me they were covered so that no one would see a skeleton in the looking glass. This didn't make me feel any better.

Since we were still waiting, I asked one of the women about the ceremony. "Is this going to be a Yuwipi ceremony?" I wondered aloud. She said, "No, it's a thanksgiving ceremony for Paul." I later found out that this was called an *Olowonpi* ceremony. All of the lights in the house would be turned out, as in a Yuwipi, making the room as dark as possible. Certainly spirits would come. I would see flashing of lights and hear spirit sounds as well as feel the spirits pass by me.

I was still a little suspicious. I wondered about all of the windows being covered and asked the women how they knew that the shaman wasn't performing conjuring tricks and fooling them into thinking that something spiritual was happening during the ceremony. They told me that so much activity takes place that if there were any people moving around they would trip over chairs, possibly impale themselves on flagsticks, and in general find it very difficult to "see" in the dark. She was convinced that Doug would bring spirits into the room.

By that time we were all assembled, or at least I thought we were. We sat and chatted with each other. As we sat there I asked what we were waiting for. They told me that the "singer" hadn't shown up yet. I wondered who or what the "singer" was. As I pondered this, Doug proceeded to open a brown suitcase he had carried with him into the room. It was quite small, so I assumed that it contained a few artifacts needed for the ceremony. I was mistaken. In the case was more than I expected it could hold. As he unloaded it, placing the various objects on the floor, I had the impression that I was watching a magician performing an ancient feat of magic. You know the kind of trick where the magician pulls an endless number of objects, which couldn't possibly fit inside, from a box on the floor.

When he had finished, the living room floor had been transformed into an altar-stage called a *Hocoka*, containing all of the ritual accouterments that he would need to perform the ceremony. The first thing he took from the case was a rug that he carefully placed on the floor. The rug was darkly colored, appearing quite black, and quite large—at least seven by ten feet. He then took a peace pipe from the case. It had a long stem and its bowl was made from red pipestone. It was roughly carved in the shape of a figure that I could not discern. He also pulled four small flags in solid colors of red, blue, yellow, and white from the case. He attached the flags to four sticks, each about three feet in length. Each color represented a spiritual force from a different direction.[10] White represented the South. Red, the North; Yellow, the East; and Blue, the West. Usually black is used instead of blue.

Later, one of the women brought in four coffee cans filled halfway with dirt.[11] Doug placed each of the miniature "flagpoles" in a separate can and placed the cans at the four corners of the dark carpet, marking off the appropriate directions of space. He also placed some eagle feathers on the rug toward the end facing the chairs in the middle line of the three-sided square.

As he continued to prepare the altar, the "singer" appeared in the doorway. His name was Marvin Red Elk. I was struck by his appearance. He was a young full-blooded Oglala Sioux with long black hair tied in the back in a ponytail. His skin was richly reddish brown, and his quite handsome face was strikingly marked by slanting but full brown eyes. He carried a large flat drum with him, at least eighteen inches to two feet in diameter, and a striker to play the drum.

[10]According to Jamie Sams, some Sioux put red in the East, yellow in the South, black in the West, and white in the North.

[11]Jamie also told me that the dirt in the coffee cans was from groundhog holes dug up by the animals. This dirt is considered to be sacred.

With the arrival of Marvin, Doug took off his shirt and his shoes and socks. He was surprisingly fit for a man who smoked like a chimney and apparently drank booze. But, as I mentioned, the booze impression was incorrect, as were most of my early impressions of him. Nevertheless, he was nearly seventy years old. I noticed that he had many wounds on his chest, mainly centered around his breasts. These were wounds given to him in the traditional Plains Indian ceremony known as the "sun dance." He also had several scars on his arms. These were from skin offerings.

He then put two large rattles on the rug. They were covered in buffalo hide. Inside the gourds were many small stones or seeds and dried pieces of his own skin. Doug then told me that no rings could be worn during the ceremony, and certainly no silver. So everyone was asked to remove any rings or silver from the hands or neck. Then a string of tiny purses, looking like small tea bags, containing tobacco and called *canli wapahte*, were strung out like popcorn strands at Christmas from flag to flag, making the area look like a small boxing ring.

When all this was in place, the closed-off area was regarded as a re-creation of the universe as seen by the Oglala Indians. The space was thus transformed from a profane or everyday state into a sacred model of the universe.

Just as we were ready to begin the ceremony, Doug, standing in the middle of the rug, turned to me and surprisingly began to joke with me. He then said, "You're going to get scared. But it is very important that you offer a prayer for Paul and me." I felt quite honored that he thought that my prayers were important. As he stood there preparing for the event I suddenly realized—with total conviction, this time—that Doug was not a drunken Indian but a powerful shaman and that all of the early experiences I had with him were just preparations and illusions to throw me off guard, to see if I was worthy enough to engage in what was about to come.

Then the lights in the room were turned out. He asked if anyone could see any light. I noticed that some light was coming in from the window near me. Doug asked me to adjust the blanket over the window. I did. He also asked anyone in the room who was wearing glasses to remove them. I removed mine. Then Marvin began chanting and playing the drum. His voice resonated throughout the room. Later on I found out that Marvin was an accomplished artist as well as a ceremonial singer. I was quite spellbound by the song he sang and felt as if I were floating.

As I sat there in the dark, I suddenly saw flashes of light coming from different places in the room. At first I suspected that Doug was striking a pocket cigarette lighter; however, I didn't hear any sound associated with these flashes. The light also didn't appear to be produced by a flashlight, either. The color of the lights was a greenish blue. The lights appeared to be like the glows from large fireflies. Then I heard the flutter of wings. The spirits had come into the room.

Later I found out that these apparitions were signals of the spirits that were summoned by Doug to help in the complete healing of Paul. The Oglala shaman operates differently than some shamans who go on a journey to the spirit world to diagnose and heal the sick. Instead the Oglala shamans call the spirits, powers from the past, into the present, and their manifestation is signaled by the flashing lights and eagle wings beating.

The Yuwipi shaman is an intermediary between the supernatural and the common people, the dead and the living, the past and the present. Through this mediation the present world is placed in contact with "all of our relations."

The term "all of my relations," *Mitak' oyas'in,* is very important. It is recited after each person has said his prayer in the ceremony.

Doug then began reciting a prayer to the six spiritual directions.[12] Later I found out that Marvin had sung this prayer in his song. Loosely translated it goes:

Friend, I will send a voice, so hear me.
Friend, I will send a voice, so hear me.
Friend, I will send a voice, so hear me.

In the West I call a black stone friend.
Friend, I will send a voice, so hear me.
Friend, I will send a voice, so hear me.

In the North I call a red stone friend.
Friend, I will send a voice, so hear me.
Friend, I will send a voice, so hear me.

In the East I call a yellow stone friend.
Friend, I will send a voice, so hear me.
Friend, I will send a voice, so hear me.

In the South I call a white stone friend.
Friend, I will send a voice, so hear me.
Friend, I will send a voice, so hear me.

[12]Jamie Sams also prayed to a seventh or inner direction.

On earth, I will call a spider friend.
Friend, I will send a voice, so hear me.
Friend, I will send a voice, so hear me.

Above, I will call a spotted eagle friend.
Friend, I will send a voice, so hear me.
Friend, I will send a voice, so hear me.

This song is considered to be the most powerful chant in the ceremony because it is directed to the entire universe. It reminded me of the three mutually perpendicular directions of space, and the dimension of time. By bringing in the spirits throughout the past, the dimension of time is brought in, so the whole four-dimensional space-time continuum is evoked in the ceremony.

After the song and the prayer by Doug, I heard violent shakings of the rattles. I nearly jumped out of my chair when suddenly one of the rattles struck the floor between my legs. I was quite shaken. I was also amazed when I realized that when the gourd struck the ground it glowed the same brilliant blue-green color as the "fireflies" I had seen earlier.

After thinking about it, I realized that the color was similar to that produced by a phosphorescent material. However, I had never seen a material phosphoresce in this manner when it was struck against another object. Perhaps it was some form of static electrical discharge. On a dry night, if you rub the folds of a nylon or other synthetic fabric blanket together you can produce sparks of color that are similar to what I saw. However, this explanation doesn't quite work for the flashes I saw, because the sparks are always accompanied by the sounds of the sparks. These colors flashed silently.

The whole ceremony continued in this manner. There were songs, chants, and prayers by Doug and the continual surprises of flashing lights and eagle wings. Quite often the flashes and wingbeats were only inches from my face and body, and each time it happened, I was startled. After about an hour, the room grew silent. Doug asked each person in the room to offer a prayer. Although I was the only white person in the room, not all of the Sioux present spoke the Sioux language. But each one offered a prayer.

It was then that I realized how seriously these people took the experience. I heard many different kinds of prayer. Some wished for better health. Others wanted to have their personal relationships improve. A few were bothered by spirits and asked if other spirits would come into their homes and clean out the bad spirits.

I was moved by the humble and open spirits of my new friends, and I felt as if I were a part of their family. A kind of relationship or feeling of relatedness took place in the room. I remember another ceremony like it. I once attended a Narcotics-Alcoholics Anonymous meeting with my son, Michael, and felt a similar presence of spiritual force when each person "confessed" and told honestly about his life and problems with drug abuse.

When my turn came, I offered a prayer. I first prayed for Doug and Paul's safe journey to Canada, as I was instructed to. I thanked "all my relations" present in the room for inviting me to partake in the experience. I then offered a prayer for better understanding and more felt love between the many peoples of the world. After each prayer, the ensemble would say "Ho," signifying the acceptance of me and my prayer. I felt that my prayer was moving to them and I believed that they took what I said to heart.

Then the ceremony was officially ended, and the lights were turned on. I then thought over the experience and realized that all of us were still seated in our chairs, and yet during the darkness, I had felt as if the room was full of moving people. With the room lights on I then looked at the altar. Everything that was laid out was completely scattered about. Some of the cans had dirt spilled out around them, and everything was in disarray.

I was astonished when I looked up at Doug. He was standing in the middle of the altar space with blood dripping from a cut just under his right eye.

Doug then quietly said to one of the women who had helped him set up the room, "You forgot to include the raw beef liver in the altar. The spirits were angry at this." He didn't offer any reason for his bleeding. But I surmised that it was a mark from the spirits because he forgot to include the liver in the ceremony.

Then Doug turned to me. He began to kid me again in a good-natured manner. He wanted to know if I was very frightened during the ceremony. I got the feeling that it was better for me to tell him the truth and say how frightened I was. That gave him a good laugh. I told him that I was very frightened and that, when the rattle banged on the floor between my legs, I nearly jumped out of my chair. He was laughing and nearly rolling on the floor. I told him that if that gourd came any closer to me, I would have jumped out of the window. He howled with laughter. He then proceeded to gather up all of the materials he had placed on the floor and put them back into his small suitcase.

Paul got up from his chair and went into the kitchen. Several deep cooking pots were placed in the middle of the floor. Each contained a typical Sioux food. One had a stew in it. Another had a jam made from chokecherries. Fried bread was offered, which tasted very much like sopapillas served in New Mexico. Jell-O was offered and Kool-Aid or coffee was served. Everything was quite tasty, and I enjoyed myself. I was surprised at how hungry I was.

As we were sitting around digesting our food, I noticed that Doug had eaten very little. Apparently he was used to fasting and felt little hunger. I then mentioned to everyone that first they frighten me and then they fatten me up. Doug laughed at that, and all of the people there joined in, in good spirit. I told him that even though the lights were now on, I still felt the presence of the spirits. Doug nodded his head in acknowledgment and enjoyed my spontaneous offering of humor and my playing the part of the fool.

Later each of them told me that he or she had been frightened the first time when going through such a ceremony. My lack of bravado was refreshing to them. Then, as it was now quite late, I went up to each person, shook his or her hand, and said thanks for having me. I said good-bye and walked outside to get to my car. Doug then said, "Be careful when you drive. Watch out for playful spirits."

Doug and Paul then followed me out. As we stood in the moonlight, Doug said that when he got back from Canada, he would take me up to a mountain for a vision quest, if I wanted it. I didn't feel very keen about the idea, because I was afraid that his Heyokah spirit would make me too upset. Perhaps I was being a "chicken." Perhaps I was being prudent. He would be to me as Don Juan was to Carlos Castañeda. He would be perhaps my "petty tyrant." I remembered how he had fooled me throughout the day, at first appearing as a drunken Indian, somewhat incoherent, and then as a powerful shaman fully in control of his physical and spiritual faculties. When I first met him, he appeared old and shrunken. When he performed the ceremony, he appeared tall and youthful.

I was about to get in my car when Doug approached me again and said, "Now when you drive home, remember to look straight ahead." He then turned away and went to a vacant space on the land, bent over at the waist and threw up what little he had consumed. I was again reminded of the drunken Indian image. Was he just sick with booze? I still doubted.

After seeing that Doug was okay, I said good-bye and got into my car, little suspecting (although I was warned) that my journey into the spirit world was not over yet. I drove carefully along the dirt road leading out to the highway threading through the reservation. I remembered to turn right to get back to the medical compound where I was staying in Pine Ridge.

I was driving along feeling quite content when I suddenly realized that I was driving in the wrong direction. I must have driven around ten miles when I felt this. I turned around and proceeded back, bemoaning that it was late and that I would have to drive at least another twenty or so miles before I would get home. I was getting quite tired.

After driving for around fifteen miles I again had the feeling I was going in the wrong direction. I was totally confused. So I turned around once again and headed back in the same direction I had originally taken when I reached the highway. Finally, after driving about an hour, a trip that should have taken me no more than fifteen minutes, I managed to make my way back to Pine Ridge and my apartment on the compound. The jokester-shaman was still playing with me.

I left the next day for New Mexico. I had finally been exposed to a Native American shamanic ritual. It was not the last I was to experience.

The movie scene jump-cut to Jorge and Miguel talking. And now, as I sit here in Lima, watching Jorge, the movie playboy in the scene, and remember that he was my real-life shaman, something fell into place for me. Doug was a trickster-shaman, and he had healed some part of me by making me face my "buttons," my irrational fears based upon physical appearances.

My hypothesis that shamans will do anything to convince the patient to change his or her belief structures snapped into my mind. A few months ago, Holger Kalweit and his wife, Amelie Schenk, shamanic researchers who live in Switzerland, told me that they observed Tibetan shamans quaffing down lighted butter lamps before their audiences. Since the butter lamps were burning when they did this, it appeared that the shamans were swallowing hot melted oil. Actually the oil was not really hot. Only the surface of the oil was burning, the major portion of the oil was below the surface and remained relatively cool.

Holger, too, swallowed down the contents of the burning butter lamp to test out his theory. By taking the liquid into his mouth and closing his mouth quickly afterward, thus quenching the fire, there was no burning and the oil was actually found not to be very hot to the tongue. Holger's story reminded me of my earlier shamanic experiment with firewalking, although I was not able to convince myself that it was all a trick.

I also realized another insight. Although shamans are very sincere, they do use gullibility and any form of trick they can to alter a patient's own consciousness so that he or she will accept the existence of a power, an archetype from the imaginal realm, greater than a patient's own mind. This makes it very confusing for an overlooker seeking chicanery. All he will see is the clever conjurer performing the act. What he cannot see is the effect that this act is having on the patient, which is to get the patient to alter his or her belief system, a la hypothesis 4. All the overlooker can see is blood bags concealed in the palm or a thumb tip of the shaman, whether or not they are there.

Now I must admit, I didn't really attempt to debunk Doug or any of the shamans I had met. I was attempting to "see" what they saw. I was willing to stretch my critical thinking and just simply "hang out" with them. I cannot say that, after all of this, I had every shamanic experience there is to have. But I had enough of a peek into their world to now believe it was genuine. It was as real to me as an electron.

But my major insight went beyond tricks. I now knew that a shaman was a kind of physicist. I knew that my initiations were necessary. I was an experiment to the shamans. I had to be prepared as carefully as any physicist prepares an experiment. However, even the best preparation will not allow the physicist to really see what is happening at the atomic and subatomic level. The uncertainty principle is inviolate. No matter how carefully you prepare a system, it will not always repeat the values expected. Electrons elude all attempts to pin them down.

There was a reason for this. Electrons are only tendencies, concepts, ideas, and probabilities. They are not as real as we picture them to be in our mechanical mind's eyes. I began to feel that electrons, too, inhabit two worlds: they are part of the physical world and they are part of the mental world. The two worlds overlap every time a physicist attempts to find one. Electrons are in some sense conscious, and when an experiment is carried out, the mind of the physicist links with the elemental mind of the electron. In a sense, this is what the observer effect is all about: linking the observed to the observer. It's all one mind at work.

A similar thing happens to one undergoing shamanic initiation. The mind of the shaman and the mind of the patient become one. This would be impossible if the patient's mind was not altered. Just as a physicist alters an electron's reality by putting it through the ordeal of experimentation and making it appear as a wave or a particle, my mind had become attuned to the shamanic mind that was investigating me.

And I was beginning to feel that my connection to the shamans was not over. Just as a physicist is connected to all electrons in any part of the world he happens to be because he knows how to invoke them, once I had undergone shamanic initiation, my mind was attuned to all the shamans that had prepared me because they knew how to invoke me. I was beginning to realize that by being a participant, I was becoming a part of the shamanic world and no longer an objective observer.

Yet, I still was in doubt of this. I had not been brought up in a shamanic tradition. Shamans were somehow created by the lands that bore them, and they were aware of the sacred plant life surrounding them and how that plant life influenced their shamanic ability.

I needed to know how the earth and its plants created shamans. I remembered what I had learned. The planet itself, the mother earth, gave birth to everything necessary to make a shaman. All one had to do was believe that it was possible.

5

Sacred Places and Plants

The movie went on. The initiation ceremony for Nexy and Miguel had ended, but Nexy was still troubled. She was haunted by spirits called *yacurunas*—phantoms arising from the Amazon River—that are actually the pink dolphins found there. The movie shaman, Meliton, tells them that he must take them to a sacred place deep into the jungle where Nexy must be given a special potion. In order for her to heal she must discover how her past is creating her illness. She is suffering from psychological problems caused by the death of her mother when she was a young child.

The depth of the jungle was a metaphor for the depth of the unconscious mind. I began to feel that there was a connection between the unconscious mind and the consciousness of the planet Earth. I suspected that the connection had to do with sacred plants, and potions made from those plants, and that sacred plants grew on sacred ground.

I began to think again about what there was in quantum physics that could indicate such a profound connection. I then recalled my idea of the hologram. You will recall that in my vision everything was connected in a holographic manner. But there is one rather exceptional property of holograms that I did not mention earlier. Since every part of the hologram contains recorded light waves from every part of the object, it turns out that each tiny portion of that hologram actually contains the image of the whole object! This can be seen when one breaks the emulsion into tiny pieces and illuminates each piece with the laser beam.

If we and everything in the universe are constructed from holographic waves, then, if my analogy was correct, each part of ourselves contains the whole universe, including the planet Earth. Perhaps my analogy meant that our consciousness was also connected in the same manner to the planet's consciousness. Perhaps being in sacred places reinforced that connection in the same manner that an image seen through a part of the hologram is more enhanced when several pieces of the broken hologram are put together like a jigsaw puzzle. This meant that merely being at a sacred place would reinforce a shaman's healing powers. That was why Meliton wanted to take Nexy and Miguel deep into the jungle to a sacred spot. And since the planet had a consciousness of its own, perhaps we could tune into that consciousness if we were to take into our systems the sacred plants that the mother earth had produced. In our culture today, this is frowned on to a large extent. Our laws make the use of sacred plants nearly impossible, so we are cut off from a sacred heritage.

As I thought about the power of sacred places and plants, I recalled the seventh hypothesis, that shamans worked with a sense of a higher power, and now I suspected that power was found in the earth itself, reinforced by ingesting sacred plants.

So perhaps it was where shamans were born that made them, giving them their healing powers. And perhaps we, too, could partake of some of these powers by being in sacred places, ingesting sacred plants.

Watching Meliton lead Nexy and Miguel into the jungle to their destiny with a sacred place, I thought again of the shamans from Brighton, England, and what they told me about the sacred condition of our mother earth. By being exposed to a sacred place and plants obtained from there, the two people were learning to become shamans themselves.

It was dawning on me that it was my belief system that kept me from seeing the truth. It was my mechanical mind still running the show. It was my belief that the planet was there to be conquered, not respected, that kept me from my discovery.

WHERE DOES SHAMANISM WORK BEST?

I was back in Brighton with the two Anglo-Saxon shamans/researchers. It was the summer of 1988. We were sitting in Richard Dufton's living room. Our conversation had shifted. We were looking at the differences in shamanic activity as they existed in different countries.

Richard asked me, "Where does the Qabala and the theories of it work best? In Judea?"

I said that it was hard to say and I didn't really know.

Then Richard said, "I look at it this way. We are the human organisms grown on a particular patch of land. We eat that land. Our mothers ate it. We are that which our mothers ate and that which our forefathers were."

Shamans do come from particular land areas. They soon recognize where the power spots are. Shamans are sensitive to power spots in their own countries because they are born of that land. I was realizing that shamans remain sensitive to their own land vibration wherever they happen to be.

Richard went on to explain that these power spots have a certain vibrational pattern, a harmonic pattern that is highly specific to the land mass present at that spot alone.

I realized this meant that the land vibrates with a specific frequency. This vibration is very subtle, and usually we are unable to detect it. During times of earthquakes, the vibration is not subtle at all and we feel it quite strongly.

I remembered the first time I noticed the earth's subtle vibration. I was visiting a laser laboratory. In order to make long-exposure holograms, film emulsions and a laser must be used. However, everything must be mounted very carefully because vibrations upset the delicate laser light and cause the image to become blurred. Consequently the whole image-making apparatus must be mounted on a special table that is able to withstand even the smallest earth tremors.

The whole planet Earth vibrates. It has a major harmonic vibration—something like eighty-eight cycles per minute. However, superimposed on that major harmonic are minor harmonics. East Indian sitars and Scottish bagpipes use this principle. The major harmonic is called the drone, and the minor harmonics carry the melody. The melody is always in relationship with the drone.

Each different land mass has its own melody—its own subharmonics. Shamanic practice or magic apparently depends on these harmonics. They subtly influence everything that is sensed.

I began to think that the earth's vibrations might be related to the subtle vibrations of the human body. Then it occurred to me that quantum physics is really a science of vibrations, subtle movements of what physicists call quantum waves of probability. Even though, as I mentioned earlier, we have never seen these waves directly, these wave patterns are necessary for us to understand the patterns of atoms in molecular configurations. All molecular patterns are the result of wave interferences. Wherever there is a wave buildup, atoms are likely to be present. Wherever the interference of these waves causes a cancellation, atoms are less likely to be present. Thus by adjusting the ways in which these waves combine, molecular structure itself is altered. Could the mind of a person cause these wave patterns to change? Was this a way in which cells got out of harmony with each other?

I then realized what the connection between a sacred place and a shaman was. Shamans were able to use their healing powers best when they were working on the very grounds from which they grew themselves. Their vibrational patterns were more closely tuned to their home grounds. Shamans have tuned to their own sacred grounds throughout history and have evolved their own particular forms of rituals and healing.

Richard said, "So the native magic, and this is real physics, works best in its native country. All magic will work everywhere. But there are some places it will excel and some places where it is spiritually useless. Qabala is useless in the Arctic Circle. There is no land there."

Chris Hall added, "Believe me, if you get a Druidic person and stick them in the middle of the Sinai, and let them go through a ritual, it will be just so much garbage. But if you put them in the middle of Stonehenge, on the solstice, you've got power."

NORSE RUNES DON'T WORK IN MEXICO

As I looked around Richard's living room, I was quite amazed at the number of magical artifacts he possessed. Each came from a different power spot. That made me think of the time when my friend Ralph Blum gave me a crystal piece engraved with runes to wear around my neck.

I said, "I have to tell you a story. Once not long ago, I was given a crystal with rune carvings on it and a gold chain. My friend told me that this would protect me."

Richard said, "Was it Ralph?" I was surprised that he knew who had given it to me, since I hadn't mentioned Ralph's name before.

I told Richard and Chris that I had been living in La Jolla at the time, and was about to journey to San Miguel de Allende, Mexico, to do a rewrite on a book project. I had told Ralph about my trip and said that I was a little nervous about living there. I mentioned that I felt a need to be protected. He then gave me the beautiful gold necklace containing a crystal with a rune in it.

I was very moved that he gave it to me. I wore it for weeks without incidence, but then a strange thing happened to me the first night my wife and I arrived in Mexico City. I usually wore it when I slept. But for some reason that night I took it off and put it on the nightstand. Suddenly during the night, I heard a loud noise and I didn't know what had happened. Now, I think that the crystal must have simply fallen off the nightstand near my bed.

When I awoke in the morning I had totally forgotten that it was there at all. It was out of my mind. But since we were leaving that day to take an early "blue train" to San Miguel, I searched the room as I usually do before I leave my hotel. I checked under the bed and the nightstand. I looked carefully through that whole room, didn't see anything, and because I had somehow forgotten that I had the crystal, I then assumed that we had everything with us.

We were on the train heading to San Miguel, and I suddenly remembered that I didn't have it. I felt a tinge of fear and wondered what had happened to it. I called the hotel in Mexico City the minute I arrived at our hotel in San Miguel. They checked the room but found nothing.

"What was going on?" I asked them.

Richard said, "You were going to a power spot. The Mexican use of power is very dynamically different from the Norse use of power. If you had brought those two together, it would not have been well for you."

Chris then said, "It was well for you that the crystal disappeared."

I blurted out, "I couldn't understand why I had taken it off that night, because I hadn't taken it off to sleep before. I had just gotten to Mexico City a few days before. But the last night I was to spend in Mexico City . . ."

Chris interrupted, "The Mexican magic, as described by Castañeda,[1] is exactly the same power as ours, but it is 180 degrees out of phase with ours. It hits head on. It's a different part of the hemisphere. You know that if you get a sharp north wind coming down toward the Gulf of Mexico, and a south wind coming up, wherever the two meet you get cyclones and hurricanes. We had one here in England, not too long ago."

I was back in the Lima movie house. And I was certainly feeling the effects of being in a different hemisphere. A cyclone was going off in my mind. While I now understood, through the story of vibrations, that in order to use their powers, shamans tune to their own sacred grounds, I still didn't understand what it was about these places and the plants that grew on them that enabled the shamans to heal.

Meliton, the movie shaman, had arrived with Nexy and Miguel at the sacred spot in the jungle. Why was that particular spot sacred? What made any place on the planet sacred? My question made me remember my visit

[1]All readers interested in finding the parallels between Carlos Castañeda's and Chris Hall's views can scrutinize Castañeda's *The Power of Silence* and substitute the word "wyrd" for Castañeda's "intent."

with Charla and Paul Devereux, who were sacred-site researchers, last year when I returned to Wales to visit them.

I met Paul and Charla at a wonderful old Welsh hotel in the center of Brecon called the Wellington. I asked Paul to explain what his work was about and how it might relate to my project on shamanic physics. He was originally an artist who found great interest in the geometry of the earth. His research dealt with ley lines—somewhat mysterious markings on the planet that are often mistaken for ancient roads. Paul told me that when ley lines were discovered, researchers explained them as indicating that ancient peoples placed some of their spiritual sites in alignments. He had performed extensive research[2] trying to show that the ley line has spiritual significance. While I was interested in their spiritual significance, I was after their physical significance and what they might have to do with shamanic healing power.

PLACES OF POWER

Then Paul told me about his latest investigations. He had just finished working on a book called *Places of Power.*[3] In it, he looked at the possibility that ancient sites had some form of energy associated with them. Earlier, he and Charla had started a program of investigation called the "Dragon Project,"[4] again to look at different forms of energy associated with ancient sites. They put out an appeal to many different kinds of investigators, including dowsers and physicists. He said, "We thought it would be a good idea to start a two-prong study of ancient sites. One was physical, what could we measure. The second one was using psychics, like dowsers and psychometrists, running them in tandem.

"A lot of what we found is bunkum. A lot is extremely wish-fulfilling on some people's part, but others have had experiences that I would say are authentic, such as people getting electric shocks from stones."

Then Paul asked me about some specific physics questions concerning

[2]*Lines on the Landscape,* with Nigel Pennick. London: Robert Hale, 1989.

[3]Paul Devereux. *Places of Power.* London: Blandford, 1990.

[4]The Dragon Project is a program of physical monitoring of curious forces at some prehistoric sites concurrent with a psychic archeological schedule using dowsers, sensitives, and biofeedback methods. The name is in reference to the ancient Chinese geomantic practice of using the dragon symbol for terrestrial currents. The program was started in 1977.

the observation of strange light phenomena, such as ball lightning, UFOs, and other unexplained apparitions. Could there be any connection between such phenomena and radioactivity? They had found key stones at a site that were so magnetic they set compasses spinning. They had also found radiation anomalies, specifically in prehistoric stone structures in Cornwall which were very radioactive. Paul believed that the ancient builders used these sites of high natural radioactivity as mind-changers to enhance ritual experience and hallucinogenic experience. In his book *Places of Power*, he listed around thirty sites in Britain and others in the United States and elsewhere, where magnetic or radioactive energy can be observed and measured.

Paul had also recorded, at some of the sites, that light phenomena can be seen. The sites and the lights occupy the same type of landscape, faults or geology. He believes that light phenomena have a natural explanation. "People have seen lights moving around at these sites of radioactivity and magnetic energy. In the past they were seen as fairies."

I began to think. Was there a connection among light phenomena, radioactivity, ley lines, sacred power spots, and the shamanic realm? Apparently, the answer was yes.

A number of researchers reported their findings concerning this connection in an article entitled "Angels, Aliens, and Archetypes," in a recent pair of issues of the journal *Revision*.[5] All of the researchers believe that the shamanic realm is "ontologically real."

Now "ontology" is a fancy word. It means the "logos" of existence. The word "onto" means being or existing. In a nutshell, these sightings, these UFO phenomena, the shamanic reality, near-death phenomena, and a host of other paranormal occurrences, perhaps not all such occurrences, but more than pure fantasy or liars would produce, are real. They exist. They are here to stay and they won't go away.

All of these experiences are therefore to be taken as data points, if we believe the theory that a mythic realm—that of the shaman—exists. The ley line researcher, Paul, had many documented reportings of sightings himself, and believes that these manifestations are related. The people that have these sightings are perhaps more sensitive in some receptors than most of us. Sensitive human brains are affected because the environment has affected them.

In support of this supposition, Paul believes that geophysics cannot be taken out of the equation. It has an effect on consciousness. Through pure

[5]"Angels, Aliens, and Archetypes." *Revision*. Volume 11, Numbers 3 and 4, Parts one and two, 1989.

experience over tens and tens of thousands of years, people have found that certain places are haunted—places that you don't want to go to or stay around. There is no rational reason for it. He believes that poltergeists and ghosts are not spirits of the dead but something else.

He said, "There are places where the crack between the worlds becomes very thin."

I began to speculate that there was a specific connection between shamanic consciousness and radiation discovered around sacred sites—where the "world-crack" was very thin. In fact it was, probably, radiation that induced shamanic visions. I had no idea how that radiation made shamans vibrationally attuned to the sacred grounds of their birthplaces. Perhaps it had something to do with the amount of radioactive carbon or other elements absorbed by the human system being slightly different in different countries. The shaman born of a particular land, ate of that land, and thereby contained a similar amount of radioactive carbon in his system. This would render her or him sensitive to place. Would shamans continue to go to places where these extraordinary events would occur?

Paul Devereux believed that they would. He said, "This is shamanic physics. This is the whole nature of geomancy. People have always been sensitive to place. It's only in the last few hundred years that we have lost that ability. Our entire lifestyle is very different from anything that has ever existed before. Now we put up steel-frame buildings on an economic basis not on a geomantic basis. God knows what is happening because of the electromagnetism that is bouncing all around."

We may have lost our ability to be sensitive to place simply because we desired to communicate with each other over vast distances and we desired to have energy and power at our fingertips. The electromagnetic spectrum has extended into our sensorium. It gives us information. It heats us. It illuminates our dark spaces. We merely flip a switch and we are comforted. But we also lose in the deal. Like muscles that become flabby from lack of use, our senses of imagination degenerate. The imaginal realm vanishes into fantasy—memories of a time past when we first brought fire down from the mountain.

I began to think again. Suppose that a tribe lived near a site of high anomalous activity. Such activity could be due to unusual radioactivity, such as rocks bearing radium or uranium or any other radioactive element. Or it could be due to rocks bearing magnetic material such as iron, nickel, or cobalt. In the latter case, cobalt does have a radioactive isotope, so it could be both radioactive and magnetic. Now suppose that some young children from the tribe wander off and play in the area where the anomalous activity exists. Perhaps after several days of playing in that area, some of the children become ill. This could be due to the influx of radioactivity or to some effect of magnetism on the body.

Even today we know that people who live near electrical power lines experience forms of illness,[6] perhaps even higher incidences of cancer and some forms of mental disorder. It is not too hard to believe that children playing near such power sites would, on occasion, become ill. Perhaps one of these children, being a little more sensitive than the others, even went through a near-death experience (the relevance of which I'll explain in Chapter 10).

And in this way a new tribal shaman appears. The shaman remembers the experience when he grows up and marks off the site as holy. For now the shaman has powers, and a power spot for the tribe has been discovered.

Paul then told me that he believed that these sites are remnants of the physics laboratories of Old World shamanism. Wherever a stone circle exists there also is a high incidence of light phenomena sightings. Such outcrops of stones appear in places where the geology is right. Right geology means, for example, where there are fault lines.

Every stone circle in England or Wales is within a mile of a surface fault line. This is not accidental, according to Paul. "You go to America and you'll find some of the key sites are on very bizarre and very rare geological situations. We can begin to reach back in time. Not that ancient people were going around in white lab-coats with fuming test tubes. They were just responding naturally to the energy balance in their environment. They used these sites like they used everything around them. I think that we are seeing the geophysical aspects of these ancient sites that the shamans used along with their psychotropic drugs and other tools."

CURIOUS ANOMALIES AT SACRED SITES

There was more to the story. He told me that not only are there ley lines, but there is also another form of spiritual geography. In these other sites it is possible to measure additional geophysical properties. Apart from magnetic and radiation anomalies and light phenomena, Devereux's investigators have picked up other strange happenings at ancient sacred sites—curious signals, for example, that are found only when using broad-band ultrasound detectors, and bizarre radio signals from individual stones and also in patches and areas around sites. He explained that they hadn't had

[6]See, for example, Becker, Robert O., and Selden, Gary. *The Body Electric*. New York: Morrow, 1985.

the resources to thoroughly explore them, but they were definitely detectable and they were anomalous.

Paul hypothesized, "So very funny things happen around the stone circles. Are some of these things the effect of modern energy—telecommunications and so on—or is it something to do with the actual place and its primordial qualities? I think that at least some of these effects are primordial and that ancient people recognized them. Probably, through countless generations, they found certain stone outcrops had certain effects. The spirits lived there, and ancient people would come there on their ritual journey during the year.

"We know that something happened during the late Neolithic period, because quite suddenly, in evolutionary terms, people began to settle near such sites. Nomadic life came to an end and people began to enhance sites and put in stones marking the sacred land. They knew an awful lot about the properties of stone and all aspects of their natural environment. We believe that we have provided a very tight argument that these artifacts were related to a spirit movement in the land."

If my speculation was correct, there was an explanation based on the sensitivity of certain tribal individuals to naturally occurring sites. Those that became most sick, probably near death, became the shamans for the tribe.

Maybe I was onto something here. Shamanhood developed because people who became instruments for detecting naturally occurring radioactivity or other anomalous physical fields became shamans. In ancient days, we didn't understand such phenomena. There were no Geiger counters, compasses, or electrical power detectors. We only had human beings. Since there was no technology to speak of, there was less pollution, so it seemed evident to me that possibly shamans were, in a very real sense, the first physicists and, at the same time, the first physics experiments. They had measured what had not been seen. They had unwittingly, perhaps, discovered radioactivity and electromagnetism. But not having the Western mind bent on analysis, they did what was natural. They saw it as unseen power and held it sacred. They weren't mistaken in that regard.

Then it occurred to me. People weren't the only living things sensitive to anomalous fields surrounding sacred sites. Was it possible that plants were also affected? I knew that many shamans used mind-altering substances in their ceremonies, and it always seemed strange to me that mind-altering plants seemed to grow in certain areas or climates but not everywhere. Was there more to this story than just soil and weather conditions? Perhaps the plants contained subtle fields which acted as reminders of the sacred sites that engendered the shamans themselves.

I focused my eyes on the screen. Nexy and Miguel were seated around the altar that Meliton had set up at the sacred spot in the jungle. He produced the mixture he made from the sacred plants and held it out for the two initiates to drink. As I watched Nexy taking in the ayahuasca purge to increase her sensitivity to herself, I again recalled my visit to England the previous year when I had first learned of the connection between sacred places and mind-altering substances found in sacred plants.

DRUGS AND EVOLUTION

My return to England in the summer, 1988, was marked by a little anxiety. Aside from the rapid growth and economic improvement I saw there, I wondered if I could make my old contacts again, especially with the Druids that I had met with when I was last in England. I had felt that there was a connection between the Druids and shamanism and I knew that the Druids' history went back quite far.

The Druids are an ancient spiritual order. Originally a Druid was one of the ancient Celtic priesthood appearing in Irish and Welsh sagas and Christian legends as magicians and wizards. Today they represent a small but influential spiritual order chiefly in England and Wales and, to a lesser degree, in other parts of the world. They were often noted in the British press during the summer solstice ceremonies taking place at Stonehenge and other sacred ground in the British Isles.

My visit with the former pendragon of the Druids, Jerome Whitney, turned out to be most informative concerning the making of a shaman. Jerome, an American living in London, happened to have been appointed as the pendragon of the Druid Order in 1981.

The Druid Order, like many modern corporations, has a hierarchy. The head of the Druid Order is called the chief. The chief is not unlike the CEO of a corporation. The administrator, the one who carries out the orders of the chief, is called the pendragon. The secretary of the order is called the scribe. To be appointed the pendragon of the order is quite a high honor for a British citizen, yet for an American cousin nearly unthinkable, I would assume.

As Jerome and I talked, it was clear to me that shamanhood can be linked to the use of mind-altering substances. Jerome didn't actually say this. But as we talked I began to see another link in the chain. Of course, shamans use something to induce a change in their consciousness. I asked Jerome about this connection.

He told me that everything is a drug. Mind-altering drugs are partially a function of the diet and the way of life that you lead. Something that to us is not mind-altering, like peppermint, would be mind-altering to a person of another culture who hadn't been accustomed to it. For example, sulfur in the atmosphere, sugar, or other substances like phosphoric acid could be mind-altering.

I remember talking with people who smoked marijuana every day. I asked them how it was to be high every day. They said they didn't know because they weren't high. They became totally immune to it.

Jerome replied, "That's right. I mentioned the sulfur dioxide because of the Oracle of Delphi at the temple of Apollo."

He told me that sulfur dioxide still comes out of the volcanically active ground at Delphi and that it has been doing this for the past five thousand years. In ancient days these fumes were ducted into the temple and were used by a woman called the Pythea. She breathed in those fumes.

The year before, I had been to the big island of Hawaii and had visited the active volcano site there. I did manage to breathe in some of those fumes, voluntarily, and I can tell you that it wasn't any fun at all. I was congested for several days afterward.

Jerome told me more. The Pythea sat on a special chair in the temple that was considered to be the seat of authority. Temple followers would then bring in people who wanted to know the future—whether they would win the war, how the crops would be, etc. They would then ask the question of the Pythea. She would, however, speak in gibberish. She was not conscious in the normal sense.

I wasn't surprised that she would speak in gibberish. I could hardly see straight after breathing in the fumes of hell from the Hawaiian volcano.

That gibberish, according to Jerome, would then be translated by the priest who spoke to the questioner. This is how the oracle spoke through the Pythea. But the language of the Pythea was not understandable to anyone except the priest. The drug-induced state was a key to the process.

It was clear to me that drugs certainly altered consciousness. It was also clear that drugs were used to help people, not cause them harm. But some drugs, such as alcohol, seemed to have caused more harm than help. Of course, I knew about moderate use of alcohol, but I suspected that alcohol might have a deeper history than I had known about. I asked Jerome.

Jerome then paused for a moment, sipped a little tea, and told me that in the process of evolving, new brain strategies had to be used. Nature and the guides of humanity—those incarnating spirits who are responsible for us—had to use strategies to get us to evolve. One of the strategies was to use alcohol.

Alcohol was used to put a barrier between the early human's inner and outer consciousness. In fact, it shut the inner consciousness off completely. This enabled early humans to reach outside of themselves to learn more. So alcohol was given to humanity as a device to shut down the inner knowing process.

Jerome explained that early humans were not able to differentiate between their inner and outer consciousness. They saw little difference between dreams and reality. He explained that, according to the Druids' ideas, consciousness existed after death. Consequently early humans, just after death, not being able to tell the difference between dreams and reality, did not know that they had died. (In Chapter 10 I'll go into this in more detail.)

I was startled to think about death in this way. Without any outer consciousness, assuming that an inner consciousness existed, how could you know if you were dead or alive? I was curious to explore what this inner consciousness was. I asked him if the inner knowing consciousness was more primitive.

He explained that it was a group consciousness. People weren't individuals. The person didn't have a concept of an "I." Instead early humans participated in a one-mind dream when they slept and even when they were awake. Today we still find remnants of this odd state of consciousness in the Australian aborigines and their participation in what they call the "dreamtime." The big problem for the early human was individual action.

I remembered reading Julian Jaynes's book, *The Origin of Consciousness in the Breakdown of the Bicameral Mind*. He, too, offered the theory that early man had no individual consciousness. I speculated in my book, *Taking the Quantum Leap*, that a special quantum physical agency had to be developed in the human brain. And that agency enabled humans to tell things apart—one from the other. I suspected that this agency was the same development that the pendragon was referring to.

AYAHUASCA HISTORY

I was quite impressed with the Druids' vision of drug history. I had never even conceived that any substance could be a drug when it was first introduced into a culture. The thought that humanity used drugs as an inducement for evolution was certainly a shock.

I had that uncanny feeling of destiny come over me again. And as I watched Nexy finishing her drink of ayahuasca, I remembered how I had decided after my meeting and early experiences with the shaman Jorge Gonzalez that I, too, needed to take the powerful hallucinogenic substance ayahuasca in Peru at a sacred site. Was I destined to take this substance into my system? Was this part of my evolution too? But what was this substance really? A few weeks earlier, just after my arrival in Peru, I decided to do a little research.[7]

I found out that ayahuasca had been discovered in archeological findings in Ecuador and that other psychotropic plants were also used as far back as 3000 B.C. I read that Ecuador has been inhabited by humankind for the past 11,000 years and there exist a plethora of drawings on pottery, from the Santa Elena peninsula, 4000–2000 B.C., showing people chewing leaves and inhaling ground plants. Drawings also show the existence of shamans. Some of the drawings appear to reveal the mental effects produced by these sacred plants.

Most of the consumption of ayahuasca took place in the Amazon region of Ecuador, which for various reasons hadn't been a focal point for study by archeologists. I wondered why and looked through some other materials.

In a study by Luis Luna,[8] I found out that shamans are also known by the term *vegetalistas*, which should not be confused with the term herbalists—persons knowledgeable in the use of medicinal plants. Vegetalistas are also versed in the use of medicinal herbs and plants, but herbalists do not use the psychotropic plants.

Another term that Luna discussed was *mestizo*. This term causes some difficulties among scholars. It means a person who has a mixed ancestry, usually of European non-Caucasian and American Indian stock. It also means a completely enculturated Central or South American Indian. Thus it has both cultural and genetic meaning, but these two categories do not necessarily overlap. Among the vegetalistas, there are both pure European stock and other ethnic groups.

Luna, himself, participated in several ayahuasca sessions.

FUNCTIONS OF AYAHUASCA

Luna categorized several important functions in the use of ayahuasca. One important function, during an ayahuasca session, is found in the shaman's *icaro* (song). Icaros are used to invoke the spirits of the plants and to invoke

[7]A major volume of work is found in *America Indigena: Instituto Indigenista Intramericano*. Vol. XLVI. Mexico: Instituto Indigenista Interamericano, Insurgentes sur No. 1690, Colonia Florida Mexico, D.F., 1986. In particular see the article by Plutarco Naranjo, "Ayahuasca in Archeological Ecuador."

[8]Luna, Luis Eduardo. *Vegetalismo: Shamanism Among the Mestizo Population of the Peruvian Amazon*. Stockholm, Sweden: The University of Stockholm, 1983.

a dead shaman or spirits. They are also used to travel to other realms and to deal with beings that exist in these realms. Icaros can modify and tune vision. For instance, hunters and healers listen to the songs sung by the shaman to help them.

Shamans can sing certain songs that provide a crossover into the visual field. The songs induce visions of geometric design. Angelika Gebhart-Sayer, an anthropologist from the University of Tübingen in Germany, wrote in Luna's study:

> The shaman's songs can, so to speak, be heard in a visual way. The geometrical may be seen acoustically. This phenomenon or crossover is often referred to in the text of the shamanic songs. A medicine person may be calling "My painting song," "My painted song," "My voice," "My little painted vessel," "My words with those designs," or "My ringing pattern."

I thought that the crossover pattern could be related to my theory[9] about the way the brain works. The ayahuasca could be affecting the neural cortex in both the aural and visual areas. Usually the sound stimulus affects mainly the aural cortex. The other cortical regions remain unaffected. But under ayahuasca, it appears that the whole cortex is excited, including the visual cortex. Consequently, a sound could act as a stimulus for visions. I wondered if I would experience any crossover with Don Solon, the shaman Jorge had told me about and whom I had come to meet.

A good Ayahuasquero (you'll remember that this term refers to shamans who use ayahuasca) will always prepare his mixture so that the imbiber becomes *mareado* (literally seasick) in order to produce the full psychotropic effect. A good brew is called *seveclarito*, meaning that the visions are clear and that they look real.

Ayahuasqueros also add ayahuasca to foods to see how their visions change, since foods left in the stomach and intestines to digest will interfere with the effects of the ayahuasca. Knowing how pure ayahuasca acts, they then compare their journey with the new substance added to the unadulterated mixture. In this way they have learned which foods are helpful and which are harmful. So ayahuasca is a teacher, and once one knows its teachings, it tells the imbiber what these other substances have to say.

Ayahuasca, like other plant teachers, is used to explore both this world and other parallel worlds that are usually beyond our normal perception. By ingesting it, the Ayahuasquero is freed from the normal space-time

[9]See my article, "On the Quantum Physical Theory of Subjective Antedating." *Journal of Theoretical Biology*. Vol. 136, pp. 13–19, 1989.

boundaries of this world and, with training, freely moves from world to world. The key factor in this is the diet. I learned that the length of the fast or specificity of the diet determines just how far the seeker will voyage.

Not all ayahuasca journeys are seen as beneficial. The Western idea of a "bad trip" is also found among vegetalistas. Sometimes the shamans attribute their bad trip to the plant, saying that it was a *purga-brava* (brave purge). Other times bad trips are caused by the person breaking the dietary or sexual restriction placed on them. According to Luna's discoveries couples should not engage in sexual activity several days before and after the ayahuasca ceremony. At other times a bad trip could be caused by evil doings of other shamans. Some shamans believe that there are persons with "weak souls." These persons are advised not to take ayahuasca. If they do, they will do crazy things, like stripping naked and running rampant through the jungle or fighting with other participants. Many native persons report that after taking the vision vine, they see themselves turned into animals, usually snakes or jaguars. Many see themselves being devoured by huge boas or having the boa enter their own mouth.

However, shamans stress that regardless of the visions, terrible or frightening, the person having them should remain as calm as possible. Even if an evil spirit appears, if the person is not afraid and remains calm, the spirit is powerless to do any harm.

AYAHUASCA RULES

Luna described the rules of respect regarding journeys with ayahuasca. Respect that:

1. Some plants and animals have powerful spirits. Knowledge and power can be derived from them.
2. To learn from these spirits, one must follow a strict diet and maintain sexual segregation.
3. These plants are tools for exploring the natural and supernatural realms.
4. There are dangers in these realms. A novice should use ayahuasca only in the presence of an experienced shaman.
5. A novice needs spirits to protect him when he journeys.
6. Dreams and visions may be manipulated by the shaman.
7. There is a substance embedded in the shaman's chest that has knowledge and power.
8. There are knowledge and power in the shaman's songs.
9. Illness is due to the actions of evil spirits or evil sorcerers.

10. The spirit is pathogenic. It has a material manifestation, such as an insect or a piece of bone.
11. These spirits can be diverted by the shaman's use of song, magic plants, and, most important, the use of tobacco smoke.

THE SHAMANIC PLANET

I now had some insights. Ayahuasca and probably other mind-altering substances were not discovered accidentally. These plants were necessary for our evolution as a species. They enabled us to reconnect with our planet. They provided us with spiritual and mythic insights into our own natures. In every culture, shamans were and still are those people most sensitive to the vibrations of the planet, and they were connectors taking their tribes outside the realms of physical existence, enabling the tribes to remember and rekindle their sacred functions. I realized that our modern culture was in danger of losing its connection with the planet, as witnessed throughout history. My early research had shown me that shamans and magic played major roles in all of the early Germanic and Anglo-Saxon tribes that lived throughout Europe. It was also clear that the Romans' use of Christianity had made every attempt to wipe out the magic and replace it by the power of Rome.

In the olden days, the Romans wanted to gain control over the many thousands of tribes that existed, scattered through the Anglo-Saxon forests. Each tribe had its own structure, its own way of surviving. And, of course, each tribe had its warrior-king and its shaman.

But when the Romans conquered they established that no one needed to believe in their warrior-king any longer. There was to be one king, the leader from Rome. And there was the one power, the one big shaman, Jesus Christ. Yes, Christ was thought of as a shaman. After all, the only model for a healer was the shaman, since there were no medical doctors around in those days.

This speculation then marks the beginning of the end of European shamanism. We can also envision how Christianity attempted to do the same thing in the New World, through, for instance, the Spanish Inquisition and the wipe-out of the American Indians.

So Christianity replaces the tribal shaman and the big Caesar replaces the smaller warrior-kings, leading to the current problems of Western civilization and the loss of our belief in individual power to heal.

But why didn't the Roman powers that swept Europe attempt to incorporate the shamanic powers that existed? Why did tribal shamanism and magic get wiped out?

By taking away the power of healing from the tribal healers and also taking away their spiritual pagan belief system, the Romans invested that power into the hands of the controller. We call this government. The key was control of nature, whether it was man, beast, or plant. And the roots for that control came right out of our Greek and Roman heritage. It was the Greeks that came up with the notion that nature could be figured out rationally. By bringing in Christianity and the Greek concept of the rationality of nature, we have the tendrils of control that were to lead to the rise of our present civilization and the destruction of human shamanic power.

But why did human nature choose this alternative?

The answer was survival. We formed central governments to enhance our abilities to survive. But we paid a dear price for centralism. We gain a stronger economy but at the price of perhaps losing our magical souls. On a global scale, we are losing our planetary soul. And drastic action must be taken. The medicine is always bitter. Whenever we have attempted to right a wrong without the use of our souls, without attempting to reconnect to our spiritual ancestors, we fight a losing battle and the situation gets worse, not better.

My mind was full. I hardly dared even think about it. I saw that my own path through this Western world was, in some real sense that I could only glimpse shadows of, paralleling the path of history itself. I began to see myself in a mythic sense as part of all humankind. As a child I was interested in magic. As a young man, I studied physics in an attempt to understand how nature worked and how humans could control her. In my studies, I had lost the magic. And now in my middle years, I am back to magic again.

I went back to thinking of the use of shamanic substances. We call these substances drugs. In our society we are quite frightened about using any form of "controlled substance" that can induce a change in consciousness. But, I then realized, if I was to really discover the shamanic world, I had to also undergo taking in shamanic substances; and to really test my ideas, I had to take them where they occurred naturally.

I now realized fully that shamans were here for an important purpose. They were here to help us reconnect with the earth, and through the use of healing plants, vision-producing or not, they were here to heal humankind. They were reminders of our ancient spiritual souls, something that is far too easy to forget in our everyday Western lives.

And I realized that even though I had chosen to find them, they had chosen to find me. The hologram image was more than a metaphor. We are connected to our planet. We are made of the planet. We are the planet.

And the sacred substance ayahuasca, as strong and as purging as it was, it was necessary for me to take it and to take it where it grew naturally—in the Peruvian Amazon.

My first two experiences with ayahuasca were powerful, but they took place in the United States. I then wondered what I would experience when I took my first ayahuasca journey in Peru. I soon found out.

6

Sexuality, Magic, and Potency

As I watched the movie, I thought about Miguel's plight. Meliton tells Nexy that she must spend some time alone. But Miguel is upset. He has fallen in love with Nexy and he doesn't want to leave her. He sees in her a mystery that symbolizes the merger of the world of the sophisticated scientist and the mystical world of the Indian from the jungle. She kindles in him a dream of his own redemption. Miguel had not realized how he felt when he followed her and Meliton, the shaman, into the depths of the jungle. He did not know that he was actually following a path into his own destiny and that Nexy was part of that destiny. Nexy is inhabited by spirits of the jungle. But to Miguel, she, herself, is becoming a spirit of the plant ayahuasca and she is the remembrance of the dark feminine intuitive self that he feels only dimly inside of himself.

In a very strange and mystical way, I had followed Miguel's path. Like Miguel, I had been invited to the Peruvian Amazon, by Jorge Gonzalez, the shaman I had met in the United States. Not only had I gone to meet Jorge, who later had led me into my own soul, but I had met the feminine spirit of ayahuasca also. As I watched Miguel's tortured looks of love as he watched the beautiful Nexy, I remembered my own journey into my sexuality. Miguel had met Nexy. I had met Nonoy, a beautiful Peruvian spirit in Tarapoto, the city where Jorge lived. I remembered how it all began when I first came to Lima, before I went into the jungle.

. . .

After leaving the United States, I arrived at Jorge Chavez Airport in Lima. It was close to midnight on Tuesday evening, June 6, 1989. I passed through customs and immigration into the main lobby. I was greeted by a flood of tourist companies, taxi drivers, and God knows what else. I managed to find a taxi driver and paid him in dollars (he didn't ask me for intis, which was their own currency) and was driven, in a car that would have been condemned by any U.S. state motor vehicle department many years ago, to my hostel, Sandy's Hostel, which was owned by relatives of Jorge Gonzalez. The hostel was in San Ysidro, a suburb of Lima surrounded by many other posh seaside suburbs, where many of the elegant hotels have moved.

Driving to the hostel through Lima, I immediately noticed how similar it was to several other Latin American cities I had visited previously. To my eyes there didn't appear to be any rhyme or reason for the multitudes of billboards, sidewalk vendors, shacks next to major buildings, and streets pockmarked with holes that the driver seemed to sense without thinking.

Lima, the capital of Peru, was the chief city of Spanish South America from its founding in 1535 until the independence of the South American republics in the early nineteenth century. It is actually built on two sides of a river, the Rimac. It lies nestled in by the mountains of the Cerro San Cristobal. Today it is a dirty city with a serious problem due to overcrowding, slums, and air pollution. Surrounding the city are the infamous *Pueblos Jovenes* (young cities, jokingly), which are shanty settlements of squatters who have migrated from the Sierra. You can still see signs of the 1970 earthquake that shook this city.

The weather was damp and not at all hot as I was presupposed to imagine. Although it was early June, the winter had come to Lima. There was a cloud covering, almost a fog over the city that the locals call *la guarua*. Later in the morning I would notice that it blots out the sun and gives the buildings a ghostly pallor that gave me the impression that I was driving in a dream.

What had the Spanish sought in this city? I had thoughts of El Dorado, a city of gold, and conquistadores, conquerors of the New World, coming here. And a new world was created, but in all of that I felt a great loss—the loss of the feeling of the land. In all of our mastery of technology, we may have lost something. It is a precious gift that maybe we can find again. It is the loss of the sacredness of place.

On Thursday, after a day's rest, I made my way to the Museum of Gold, fashionably placed in the Monterico section of Lima in the hills above the city. As I entered I found a line of young students who, in the bravery of the many, were whistling and calling attention to my presence. Just another *gringo* to amuse the young at heart. One of them wanted my shoes. Another, a young girl, acting quite flirtatious, asked me my name, in English. When I told her, she called me "Freddie." I felt a sudden tingle when she called my name. Only people who knew me intimately ever called me by that name.

The museum's top floor was filled with instruments of war, especially armor, swords, rifles, and blunderbusses, from all different parts of the world. I wondered if this wasn't some indication of the violent underbelly of this world. I soon made my way downstairs to the exhibition of gold. It was interesting, but what caught my eye was the great number of pornographic pieces of pottery. All sorts of bowls, vessels, and pipes, which date back through the Inca period, were on exhibition. They showed every manner of sexual union possible. The most prevalent was a smiling Inca god with a huge erection proudly being displayed.

Why? What purpose did these pieces serve? I noticed that a great number of them indicated the man-on-top, woman-on-bottom pose. Aside from our conditioned sense of normality about this position, did this indicate some sort of slavery aspect? I am still not sure, and historians, who suspect this meaning, don't seem to know either.

It is written that there was an age when the gods were able to appear in physical form. Apparently artisans of that time captured this and expressed it in pottery. The gods seemed to be preoccupied, as I suppose many of us are today, with sex. They found sex not only as a fount for life but as a sublime and spiritual form of enjoyment. I believe that these symbols were used by shamans who probably used ayahuasca and that ayahuasca was probably used for thousands of years in Peru and in other neighboring countries.

I returned to my hostel around five that evening and settled in. I was leaving for the jungle to meet Jorge, the Peruvian shaman, the next morning. As I lay in bed I wondered why I had felt these definite rushes when the little girl called my name. Was this an introduction to another aspect of my shamanic education that I hadn't bargained for?

I began to gather my thoughts. What had I learned so far? The major "atoms" of shamanism seemed to be vibration, truth, and power. Through vibration, entrainment with the patient, a shaman was able to feel the truth of a patient's illness. I then realized a fourth "atom." It was love. I had felt it when Jorge had worked with me during my vision vine session with him. I had certainly felt the love of all of the shamans I had undergone ceremonies with. I was beginning to realize that, through love, a patient was able to heal by incorporating into himself a recognition of his separate parts. Was there a connection between vibration and love? I then suspected that love was a form of vibrational energy and that there was indeed a "physics of love" that the shamans had mastered.

I remembered a scene that took place nearly twenty years ago. I was in Chandigarh, India, giving a lecture on the campus of the university there. I walked by a temple, and written just above the front entrance were the words "God is Truth."

I had certainly heard "God is Love" before. I was surprised to realize that perhaps truth and love were maybe the same thing. Was truth also a vibrational pattern? The words "ring of truth" ran through my head. Truth is a felt vibrational pattern. When something is true it "rings" true. When a horseshoe fitter tests a horseshoe, he strikes it to hear the ring. Bad metal rings falsely. Although we had used these words as metaphors, I began to feel a "ring" of truth in them.

All fine and good. Truth and love gave the shaman power. But what about the patient? What kept a person in a chronic state of illness? I saw the ill body as composed of separated parts surrounded by walls. Somehow those parts were not being recognized in the body. Through some form of self-destructive urge, hate, or avoidance, these separate parts would not communicate with each other. They would be out of vibrational synchronization. If they could get back in sync, this recognition would enable the illness of the part to be healed. The whole body would then be able to communicate in vibrational resonance with the ill part. This was when the healing occurred, an act of love.

I had felt some insight. I believed, although I had no other basis for this belief than my own mind, that the key to grasping how love arises and falls, how hate occurs and vanishes, lies in the quantum physical behavior of the subatomic matter that composes us. Ancient spiritual teachings were that a battle exists between matter and light, material and spirit. I saw that this ancient vision of the fight between light and dark, matter and spirit, good and evil can be related to the discoveries of quantum physics.

Although I was in danger of over-anthropomorphizing, I knew that subatomic electrons tend to exclude other electrons and take isolated positions away from each other in the atoms and molecules that make up our bodies and minds. You will recall from Chapter 1, this "Pauli exclusion" principle (described by the physicist Wolfgang Pauli) was reflected in the quantum wave behavior of matter. Thus it affected just how electrons would behave in matter.

But what interested me was the quantum wave pattern that described a pair of electrons. If one examined that pattern, one would find that the pattern would cancel itself out wherever two electrons came close to each other. This meant that the probability of finding two electrons together in the same state was zero.

If one attempted to picture this pattern, it might look like two cyclones coming together and clashing in the region between them. Two electrons created this pattern of avoidance or, if you will in anthropomorphic terms, "hate." This tiny electronic exclusion arising out of fundamental hate or avoidance could be the ultimate cause of illness.

Yet without this pattern of exclusion, all electrons would tend to form tight orbits about their respective atomic nuclei, thus making chemistry and life impossible. We needed electronic exclusion just to breathe. We needed it to form all of the complex molecular structures necessary for life.

But since all electrons in matter behaved this way, what would produce healing and, for that matter, love? I knew that particles of light, called photons, tended to the opposite behavior. They tended to enter the same space together. Instead of avoiding each other, they exhibited a tendency to cohere. This behavior was manifested in the light produced by lasers.

Of course, ancient spiritual teachings were that light is love. Was this more than a metaphor? If so, then we could think of each subatomic photon or light particle that is transmitted through our brain and nervous system as containing a unit of love, which appears as a principle of "inclusion." This photonic inclusion principle causes photons to become indistinguishable from each other, allowing them to enter into a grand state of great power even though the photons exist in different locations throughout the brain and nervous system.

I speculated that this state produces the feeling of well-being and love and causes each human being to feel attraction to other human beings and create relationships. We feel love and compassion by simply having each person become conscious of the light within oneself.

There are two "forces" acting within us. The "quantum" force of electron exclusion tends to keep things separated. The "quantum" force of photon inclusion tends to bring things together. Between these two "forces" of exclusion (which enables atoms to form all of the molecular structures needed for life) and inclusion (which allows atoms and molecules to communicate with each other and vibrate sympathetically), human life exists. The human psyche—mind and its psychological function—exists.

I looked out of my hostel window at the yellow light across the street. The fog had set in. I remembered the pornographic pottery at the museum. I laughed at myself. Sex? Perhaps something very unexpected was to happen to me the next night in Tarapoto, the first jungle city I was to visit in Peru. Jorge was going to meet me. I was sure that I would undergo another ayahuasca ceremony, my first in Peru. But what did sex have to do with this?

I then speculated that sexual behavior lay on that boundary between love and hate, between electronic exclusion and photonic inclusion. When sexual feeling arose, a tension between these forces was felt. When people felt sexually attracted to each other, a form of healing was taking place. Sex took us out of ourselves. It enabled our consciousness to become more aware within us of the light than the material electrons. In this manner we would feel the vibrational healing of love.

I had felt another tingle as I walked through the museum. Was there some part of my own sexuality that I needed to explore? I hadn't even thought about sex until that moment and from that moment on it seemed to weave a play on my consciousness that I hadn't expected or even desired. Or so I thought, then.

Now, several weeks later, in the Lima movie house, watching the adventures of two people falling in love, I thought again about the museum and the insights I had that inside of us there are two nearly opposite forces: the healing energy of love and the destructive energy of hate. In quantum physics it is easy to see how the forces of inclusion, as displayed by light, and the forces of exclusion, as exhibited by matter, are related. They are vibrational patterns in space, just undulations.

I pictured the undulatory patterns surrounding people like auric fields. Where those patterns canceled out, there was hate. Where those patterns reinforced each other, there was love. But what happens when those patterns changed? Was sexual energy just a change in pattern from hate into love? Perhaps these were more closely related than I had thought. I knew that sexual energy, the energy of creativity, could actually heal a person. The Tantric traditions of India certainly contained such references.

Later that evening I reformulated my thoughts into hypotheses. The eighth hypothesis was that shamans used love and sexual energy as healing energy. I now believed that there was a connection between the love/healing energy of the shamans and quantum physics.

My first thoughts were of the Chumash shamans, Kote Lotah and his wife, Lin A-lul'Koy Lotah. They had told me about the differences between male and female medicine. Both were necessary for a healing to occur.

MALE AND FEMALE MEDICINE

I had met them earlier in 1989 when Judith and I made our way to California from our home in Santa Fe. We stopped off briefly for a visit with our children and friends in San Diego and then went up Highway 101 to Ventura, where the two Chumash Indian shamans lived. According to their traditions, the Chumash have a major responsibility for the western part of our continent. They are the "keepers of the western gate." This gate is located at Point Conception in California. Through this gate, the souls of the Chumash must pass when they die.

I asked them to explain the difference between female medicine and male medicine. They saw the two as totally different. They saw them in terms of plus and minus.

Simply put, male-female energy produces new growth. Kote said, "If A-lul'Koy was doing male work and I was doing male work, there would be no feminine input into it. That can't happen. In our system the women are the highest beings in the culture. We come from a matriarchal tradition. That is pretty important. Only the women and mother earth can produce. By having a matriarchal system we are tapping into that power of reproduction. This is true on all levels of consciousness, the earth and everything."

"This doesn't make the male any less important or less effective. It is just a form of harmony that males don't put out," A-lul'Koy said.

I then had another insight. The feminine energy is harmonious, wavelike in action, vibrational, and reverberatory. The male energy is pointed, particlelike, thrusting, forceful, direct, and at the moment. The female is eternal. The male is at this instant. When I told them this they agreed. My quantum physics insight fit into the Chumash philosophy. Let me explain.

WAVE AND PARTICLE ACTION

In quantum physics we deal with the universe as constructed from two movements. We call them wave and particle actions. If you will recall, I discussed them briefly in Chapter 1. Before there is any manifestation, there is only the female, the wave. In order for manifestation to occur, there must be particle action in which an object suddenly manifests, for a brief instant. That action is male—a sudden penetration of the wave, producing manifestation. That ongoing manifestation, the dance between the wave and the particle, is the continual creation of the universe.

Before my work with shamans, I hadn't really thought of the wave-particle duality as having anything to do with sexual energy. But I felt a flash of intuition nevertheless. I then remembered the teaching of Carlo Suarés. The universe was created from three "mother" letters, aleph, mem, and sheen, according to the Qabala. Aleph represented spiritual power, mem represented consciousness, and sheen was the wavelike action—the flow from aleph into consciousness. The Hebrew letter, seen, another form of the letter sheen, was the particlelike action—the flow from mem back to aleph, thus completing the circuit. The game was to achieve balance. Was that balance the same between the sexes? Was this another indication of how healing occurs—a person must be balanced between his malelike and femalelike energies, in spite of the fact that each of us is definitely one sex or the other?

Something was dawning on me. Sexual energy and healing energy were the same thing. They represented a balancing act between light and matter. I began to feel that if a human being was not able to achieve a balance between his or her forces of love and hate, light and matter, that person would become ill. Few of us are able to do this alone. Thus falling in love and feeling sexual energy is the way that most of us achieve this balance.

A SEXUAL FLIGHT ON THE WINDS OF AYAHUASCA

And I was feeling quite out of balance as I watched the film. Miguel thinks that he has fallen in love with Nexy. He actually hasn't really done this. He only believes he has because Nexy has awakened him to his feminine self. Miguel's feelings were echoing my own. They reminded me of Nonoy and how she awakened me to my feminine self.

In the old myth it is the prince that awakens the princess. Little did I know that the myth referred to the prince and princess living in each of us. Little did I suspect this when I arrived in Tarapoto, just a few weeks ago.

On Friday morning, just three days after arriving in Lima, I flew to Tarapoto in the Northern Highlands region of Peru. My first flight in Peru, as all of my later flights, had been delayed in Lima for nearly four hours. Tarapoto lies down the Mayo River valley on the very edge of the eastern Andean foothills, 356 meters (1168 feet) above the sea. It is situated in the Department of San Martin (a department in Peru means the same thing as a state in the United States) and is the home of the relatively new University of San Martin. But don't let its altitude fool you. Tarapoto is in the jungle.

I was met at the landing field (I couldn't really call it an airport, but that is what it was) by Jorge Gonzalez and Raul Espiritos, who were both professors at the university. Jorge taught in the education department and Raul taught biochemistry. Raul acted as my interpreter, since Jorge, as you'll recall, did not speak English and my Spanish was still quite poor.

Raul was quite a charming man. He smiled often, and I'm sure that he had his way with the ladies. The ladies were definitely on my mind, but I couldn't see then why nor how I was to involve myself sexually here in Peru.

Jorge told me, through Raul, that I was to partake in a ceremony that very evening. I was surprised, but yet I suspected that it was to happen that soon.

As we made our way toward Jorge's home, I was struck by the heat and the humidity. I was in an area of Peru called the "eyebrow of the jungle." It is here where the jungle actually begins. East of us, the great jungle plain gently slopes down toward the sea. West of us, the long sharp climb into the Andes. Tarapoto rests on the "eyebrow."

As soon as we arrived, I deposited my things in Jorge's humble *casa*. It was situated on a dirt road not far from the center of town and the new university. I had expected to spend the next few days in Tarapoto with Jorge. But he told me that he was leaving the very next day, Saturday, for Iquitos and wanted me to come with him. I told him that I would, even though I had been scheduled to leave for Iquitos five days later. But then he repeated to me that I was to experience my first ayahuasca ceremony in Peru that night! Knowing what it would probably do to me, I said that I would rather change my flight plans and leave for Iquitos on Monday, meeting him there. Jorge agreed and took my ticket to change it at the Aeroperu office.

Jorge and Raul then left to go to the Aeroperu office and I had a look at his house. I looked around at the many artifacts in Jorge's home. He must have had on shelves at least twenty different Peruvian artifacts and whistling pots representing something to do with ancient pre-Inca Peruvian civilizations. He had dug them up himself in an archeological dig. He told me earlier that most of them were quite valuable. It was amazing that, like the pots I found in Lima, most of them showed couples in sexual union. Usually they showed the woman servicing the man in some way, either orally or by hand.

I walked outside and noticed that, except for the few houses on the dirt road, I was right smack in the jungle. I remembered the dream I had, several months ago. I was in Santa Fe and at that time had no idea that I would be here in a jungle of Peru. My dream came back to me.

I was walking through the jungle underbrush. There was a clearing up ahead. I was being led by someone to a region that was in the deep forest. There was no light coming through the overgrowth. Suddenly up ahead, in the slight clearing, I saw them. They appeared to be like a swarm of snakes, at first. But as I got closer, I saw that they were really a swarm of people all lying down on the ground in the grassy area. They were completely naked and their bodies were covered in shiny oil reflecting what little moonlight came through the jungle canopy.

They were of various ages, sizes, and, as far as I can remember, colors. They were also of both sexes. They were young and old, plain and beautiful, and common and extraordinary. As I looked at them, they seemed to change into writhing animals, but when I looked again they were still human beings. I kept thinking I was watching a bed of snakes or maybe worms.

I noticed how sensual they all were, changing from one person to the next in what seemed to be sexual liaisons lasting any time from a few moments to longer. Their sensual oily fluid seemed to be oozing odoriferously. I was feeling myself become aroused sexually.

The light was very dim but I could still see. I watched fascinated and felt both repulsed from and attracted to what I saw. All of them were writhing and moaning in sexual pleasure. A beautiful woman in their midst then beckoned to me. Her dark head rose like a serpent's coming out of a basket of snakes. I felt hypnotized by her sensual movements. Yet, I was afraid and reluctant to join them.

Slowly I moved toward the group, and their activity seemed to increase as I did. I reached the edge of their circle, and moving to my knees, I descended and let myself fall into the group. Their bodies slithering around me felt wonderful and reassuring. Yet I still felt fear.

I then found myself caught and, it appeared to me, held by them all. Seemingly millions of appendages, arms and legs, were entwining into my own. I could not move. I felt trapped by the sexual energy of the group. I finally managed to free one of my arms from the throng. It took a considerable effort to do so. They were holding me quite tightly.

I freed my other arm and slowly began crawling to the edge of the circle. I moved across the edge and out of the ring. I turned back and watched. The attractive woman smiled at me. She looked at me very sensuously. She asked me if there was anything I wanted. I felt embarrassed, but I managed to overcome my prudish feelings and said, "Yes. I want you to suck me."

She smiled, delighted that I said this, delighted that I wasn't shy in telling her just what I wanted. She then moved gracefully through the writhing bodies.

She crawled, not walked to me. She then attached her mouth to my penis, literally, as if she were a snake. I found this exciting and was fully erect. She began to give me what I wanted, but to my surprise it didn't feel good at all. My penis relaxed. She gradually changed form and became a snake.

And now, in the middle of the movie, I realized that my dream, a few months earlier, was perhaps a portent of what was to happen to me that night.

WILFREDO AND THE SEXUAL POTION

Soon Jorge and Raul returned. Jorge told me that he wanted me to meet with another shaman who lived very close by. After a brief rest, Raul, Jorge, and I left Jorge's house.

We walked along through the neighborhood, passing a number of brick and stucco houses on the way. Many of these houses had thatched roofs made from coconut palm leaves. The heat was beginning to affect me and I felt tired.

Soon we arrived at another tiny, dark, glassless-windowed casa. I was immediately introduced to the neighborhood shaman, Wilfredo, and his family. Ironically they were all watching television and viewing a documentary about Peruvian shamans!

I shook hands with Wilfredo and we moved to a back room. At first I had difficulty understanding him, but Raul was there acting as my translator. Later I found that I was understanding him quite well. I was surprised, since I hadn't spoken any Spanish since my stay in Mexico two years ago. Had my Spanish improved that much?

As we talked, Wilfredo told me many things. He told me that he was sixty-six years old and had been practicing shamanism since 1962. He explained how he became a shaman, and he explained how he was now working with alcoholics and had been successful in curing them. This was the second time I had heard about Peruvian shamans curing alcoholism. As you will recall, I mentioned in Chapter 4 that Jorge also had told me that he too had a program for alcoholism, using ayahuasca. Wilfredo also helped alcoholics by using plants that were exceptionally visionary. He told me about which of the plants were best in producing visions and how he made his "brew."

He then looked at me very strangely, shook his head, laughed and said, "I want to prepare for you a special drink. It's a *pulga*. It is not ayahuasca.

All of the shamans use this name pulga. When people receive this kind of remedy, this pulga, it will help them. This will cure pain."

"I don't have any pain," I said.

"Never mind. This also has sexual power," Wilfredo replied.

I smiled. I remembered the Museum of Gold in Lima and all of the sexual pottery, and I recalled my dream. "Oh, that would be good." We all laughed at my remark.

I continued, "But I have no woman to be with now. Will it be good later?"

We laughed again. Wilfredo offered me a drink of the pulga. He told me that I wouldn't be able to sleep all night. Since I was going to be taking ayahuasca that night, and I certainly didn't expect to be using my sexual energy, I was somewhat reluctant to drink the mixture. He repeated that he was sixty-six years old and he was very strong, quite able to enjoy a full hard-on and a happy sexual life.

Again I wondered why the sexual connotation? Later I was to learn that these Ayahuasqueros from early times, 500 B.C. to 1500 A.D., used ayahuasca for sexual potency. In fact there are many ceramic figures showing priests with full hard-ons from that time period, especially from the Tolita culture, indicating that ayahuasca was used in phallic cult activities.

Jorge asked him what this mixture contained.

He said, "It contains only roots from a number of plants and it also contains honey. And it contains *sangria de grotto*."

At first Jorge said no to my drinking the mixture. Perhaps he was worried about me taking it and then taking ayahuasca later that evening. But then he reconsidered. I drank a shot of it. Jorge took two shots. It was very strong and bitter, with the consistency of prune juice, and it tasted alcoholic. The liquid was brown like the ayahuasca I had before and, I found out, did contain alcohol fermented from sugarcane. I made a big face after drinking it, and everyone laughed at my discomfort. I asked if it would give me visions. He told me that it wouldn't.

The old shaman wiped his lips and looked even stronger than he had before. He said, "My last wife was only twenty-two years old. My last child is four months old. I am still very strong."

We all stood up after that. He told me that he was very happy that I came to his home. As I walked out, he gave me some of the mixture in an old Coke bottle. It was for my stay in Peru. He then offered to lead me in an ayahuasca ceremony. He asked me for a copy of my book, and I told him that I would give Jorge a copy for him when Jorge returned to the United States. We walked out of the shaman's back room into the sitting room. There were now six patients waiting for him all watching television with Wilfredo's family. They looked at me curiously, smiled, and remained seated waiting for their turn.

We walked out into the sun and returned to Jorge's home. Jorge told me that two other Americans were joining us. Since they do not want to be identified, I will call them Ted and Lisa, not their real names. I had met them in New Mexico. Ted was a psychologist and Lisa, his wife, was a fellow researcher. It was fortuitous that he was there. Ted spoke Spanish fluently and, after he arrived at Jorge's home, agreed to translate any difficulties I might have in understanding any Spanish spoken during the evening's ceremony. I asked him where he had learned Spanish. He told me that he was born in Lima, Peru, but left when he was a preteenager. Raul, seeing I was in good hands, said good-bye and returned to the university.

I was tired and, after looking at the ayahuasca and *toe*[1] plants growing in the back of Jorge's home, I took a nap.

I awoke at around 5:30 and, after a short visit to the "campus" (really a series of huts and cement block buildings on a cleared ground), returned to Jorge's home.

Jorge had invited another, younger shaman, Jose Campos, to conduct the ceremony with him. As the sky grew darker, I noticed that it was about 8:15 in the evening. Just then Jose, the young Peruvian shaman, arrived with a group of friends, one of them a rather pretty young Peruvian woman. He introduced her as Nonoy. I was totally struck by her beauty and innocence. She was probably nineteen or twenty years old.

"*Vamos*, we are going," Jorge announced. We all grabbed various things for our short walk into the surrounding forest before embarking along a pathway to the rear of the houses in Jorge's neighborhood. I noticed Nonoy stealing glances at me and smiling. I smiled back.

I must have been nervous, for I also had a splitting headache. This was probably due to the sudden change in climate, passing from the cold of Lima to the heat of Tarapoto.

The half-silvered moon overhead lit the ground. We were all walking along in single file or in pairs as some of us caught up with the others. I enjoyed watching the stars of the Southern Hemisphere overhead as we moved along. The Southern Cross was quite apparent. (Did this have some effect on the Indians when the Spanish conquerors came? Did the conquerors point to the sky at the cross of Jesus?)

[1]*Toe* (pronounced toe-ā) is another plant used by shamans in healing ceremonies.

NO NO, NONOY

We soon arrived at the grounds where the ceremony was to take place. We had carried portable cushions and mattresses with us and placed them on the ground. Jorge and Jose set up the altar and brought out the potions of ayahuasca and other libations to be used in the ceremony. We all sat in a semicircle surrounding the altar, where the shamans prepared their concoctions, and I was delightfully surprised when Nonoy shared my pad and sat next to me. Her name reminded me of something I couldn't place. By that simple act of placing herself next to me, I was to find myself on a journey I had hardly expected.

The ceremony began. This was the beginning of my third venture with the vision vine. On the ground were what I recognized as the usual accouterments consisting of various bottles, tobacco, pipes, a large "squirt" bottle containing the ayahuasca mixture, and various trinkets offered by the participants to be blessed. Nonoy was the first in line and she was offered the potion. She drank it without a hitch and sat down. I wondered if she had taken it before. I was next. I walked up to the altar, kneeled down, extended my hand, and received the sharp, thick, brown liquid in a small cup.

I drank it quickly. As soon as it passed over my tongue I knew I was in trouble. It tasted much more vile than I had remembered from my previous "trips." I could barely down the brew. I got up shaking my head and returned to my place next to Nonoy. She was smiling at me and began to tease me about my broad-brimmed hat, calling me a "cowboy." (Ayahuasca and sex flashed through my mind.)

Soon all of the participants had taken their turn. We were all sitting in the semicircle, waiting for the effect. A dog in the background was barking at unseen spirits, and we were all beginning to relax. As you'll remember, ayahuasca is a purgative. Bowls for throwing up were carefully placed in front of each of us.

The ceremony surprised me a little, differing from the two other ceremonies I had experienced earlier in the United States. There the shaman used a perfumy liquid that he spread on my body during a healing ritual, using his hands. Here the shamans took the liquid in their mouths and spit it out at you. It appeared to contain alcohol, and when it struck your skin, it did feel cool. Then they would open your shirt and spit it over your chest, and then down the back of your neck. I was laughing to myself as the shamans spat on me, wondering just how most Americans would take to this experience.

Softly, at first barely perceptibly, Jose began to chant the now familiar song of the Ayahuasquero. It soon became apparent to me that Jose as well as Jorge were probably mestizos, half-Spanish and half-Indian. They were mixtures and the potions they concocted were also mixtures. (Perhaps this was the reason for the purge. It was a battle going on inside of them between their Spanish and Indian heritages.)

I held my knees to my chest and listened to the song. Soon the dizziness came upon me. Jose was chanting a song I now remembered. The words were simple. Noh, Noh, Noh, Noy, Noy; Noh, Noh, Noh, Noy, Noy, and so on. A simple rhythmic chant. (What was he saying, No, No, No, Nonoy?)

He then called on the spirits of the plants.

By this time I was getting quite mareado (literally seasick, but to Ayahuasqueros, you'll recall, it meant the substance was taking its effect). I closed my eyes and my first vision came. I saw that I was moving at extremely high speed through a field of gigantic mushrooms, which appeared to be as big as trees. I then had a series of passing visions, but I soon realized a fundamental truth of using ayahuasca: When you are tired, you have no ability to concentrate on the visions. I was unable, for the most part, to clarify my visions or get very much from the experience.

After about one hour or so, I began to feel uneasy in my stomach and threw up. I then felt tired and decided not to fight keeping my body in an uncomfortable seated position and I simply lay down on the cushion. Shortly after that I noticed Nonoy lying down next to me. Our bodies were touching and I was feeling quite strange and strongly attracted to her.

During this time I had more visions. These visions were very similar to the hypnagogic images that occur just before you fall asleep. A ramble of images passed very quickly before my mind's eye. However, under the ayahuasca I was not about to fall asleep. But I couldn't concentrate on the images and so I lost them from memory. None of them seemed to make any sense to me.

Nonoy was lying quite motionless next to me, but our bodies were still touching. I noticed that as the evening wore on, the temperature had dropped by at least ten degrees Fahrenheit, and I began to feel extremely chilled. I moved ever closer to Nonoy. She didn't move away.

It must have reached the mid-sixties by the middle of the night. I was still suffering from a bad headache, which did not seem to subside and was distracting me. I told Jose, who was conducting the ceremony, about my headache. After spitting some more perfumed water on my head and blowing tobacco smoke down my neck, he told me not to worry about it, that he would fix my head as soon as we left the ceremony.

The chanting continued as I lay there under the thick canopy in the jungle. I couldn't keep my mind off Nonoy. I wanted to snuggle up to her, holding her in my arms. But I didn't. I hardly had exchanged more than a few words with her, and the headache was still there. Many people were throwing up at various times through the ceremony.

THE SINGING JUNGLE

As I lay there, I began to notice a new experience. All through the night I had heard the barking of the dogs, but as they eventually quieted down, I noticed a new sound, one that I certainly was not familiar with. The jungle was singing. The sound grew deafening in my ears. Had anyone else been listening? Perhaps this experience was an aural hallucination. It wasn't like the crossover of senses that I mentioned in Chapter 5. I hadn't realized before that an aural hallucination was possible.

All during the ceremony, Jose frequently walked around us chanting. He apparently had come between Nonoy and myself several times, and had rung a tiny Tibetan gong by striking it with a small padded stick. It reverberated in my ears quite strongly, ringing for what seemed like many minutes even though it was quite apparent that he was no longer striking the gong.

After a while I "heard" the chime surrounding me throughout the whole jungle. I listened carefully and brought as much of my consciousness to bear on the sound as I could. It was as if the jungle was chiming the exact note that Jose had struck. Did the jungle hear the chime? Could it echo it back in some way? Soon I realized that what I heard was crickets. The chime note of that small gong somehow matched the singing-rubbing of the crickets' legs.

The ceremony ended. I was somewhat disappointed and still felt woozy, and my head was about to burst. We all stood up and stretched. I watched Nonoy. Jose had his arm around her and was comforting her. I thought that they were lovers.

We managed to make our way back to the casa de Jorge. I told Jose again about my headache. He asked me to sit down in a chair and began a "kissing-sucking-healing" ceremony with me that would have made me laugh if I hadn't felt so sick.

He took some of the perfumy liquid that he carried with him in his mouth and spat it on my head. He then put his lips on the top of my head as if to kiss me there, then sucked on my head, drawing the liquid back up into his mouth. He turned to the side and spat it out on the ground. He repeated this specifically at a place on my forehead, and then around my sinus regions at my brows. It was apparent to me that he was attempting to suck the pain from my head. Between the kissings and suckings, he chanted his healing song.

I don't know if this helped me or not. I was a little shocked and amused when Jorge gave me a couple of aspirins just after Jose had finished. But at any rate my headache began to vanish.

Of course, by now it was early Saturday morning. Jorge then reminded me that he, Ted, and Lisa were leaving for Iquitos that afternoon. I reaffirmed that I wouldn't be up to the trip and would join them there on Monday. I was really weary, and while the others began leaving to return to their homes, I flopped on the bed. I put ear plugs in my ears so that I would be able to sleep undisturbed while Jorge, Ted, and Lisa made their preparations for departure. I was soon fast asleep.

RAW FISH IN THE MORNING

When I awoke, it was already 11:00 A.M. The heat of the day was upon me. Raul was there, smiling down at me as I lay on the bed. Jorge was gone, presumably checking on the flight reservations and attending to school matters. I told Raul about the ceremony and how I felt. He smiled at me and said, "I know just what you need." He took me to a small restaurant. I rode there on the back of his motorcycle and felt better as the cooling morning air hit my skin. Raul knew the owner and asked him to bring us *cebiche* (a mixture containing onions and pickled raw fish), which Raul promised would help my headache. I also ordered a tortilla, thinking that it was a flat cornmeal pancake as served in Mexico. Wrong! It was an omelet, fried in heavy coconut oil. It was tasty, but not at all as I expected.

Still feeling quite drained, I asked Raul to take me back to the casa de Jorge. I needed more rest. Raul deposited me, saying that he would return later in the evening just as the sun went down. I made my way to a bed and lay down. The heat of the day was now coming on in full force. I wondered just how I would be able to survive it. I was still unacclimatized, feeling sick, and now overheated. I noticed an electric fan in the bedroom. I turned it on, blowing it on my body, and took a wet towel and placed it over my head. I felt wet and comfortable, and managed to fall asleep for a few hours.

I awoke around five. A slight rain had fallen and the air was slightly cooler. The sun had reached the horizon, and with the coolness I was definitely feeling more like myself. I was waiting for Raul to come. It was *Sabado noche* (Saturday night) and that meant fiesta time. Indeed there was a party that night complete with a rock band, salsa, and even entertainment.

Raul returned around 9:30 that evening to take me to the *Sabado noche* neighborhood party, which was being held in a large community compound at the center of a group of new houses recently built to house middle- and upper-class Tarapotans, particularly professors and staff at the university.

The band was quite good, playing a mixture of modern Latin music, salsa, and even American rock and roll. There was even a humorous clown character who jumped up from time to time on and off the stage, and then moved throughout the dancers. He carried a portable "mike," sang, and acted up, sometimes appearing as a female and sometimes as a male. Most of the people enjoyed his antics.

I was reminded of the Hopi ceremonies and the appearances there of the clowns dressed in black and white stripes. These clowns had sacred missions: to remind everyone of the temporary and often humorous undertone behind every serious aspect of life.

Nevertheless, I was also preoccupied. I was looking around for Nonoy. Tarapoto is a small town and I surely expected to see her there. But as the evening drew on, she was nowhere to be seen. I wished I had told her about the dance.

Soon, I left Raul to return to the casa de Jorge. I assumed I knew how to find my way back, but I was wrong. Within minutes I had become lost and found myself wandering the back dirt roads of Jorge's neighborhood.

I arrived forty-five minutes later, a trip that should have taken me only ten. The last time I was so lost I had been on the Oglala Sioux Indian reservation at Pine Ridge, South Dakota. (Was this an indication that I was with the mischievous spirits once again?)

It was around one in the morning, and I stepped outside to marvel at the cool night sky, again reassuring myself by finding the Southern Cross. I went back inside and placed mosquito netting over the bed, ear plugs in my ears to keep out the sounds of the rooster, forever crowing (who said that roosters only crow at dawn?), and turned off the light. Soon I was asleep.

NIGHT WINDS OF AYAHUASCA

It was the next day. Jose Campos had told me that he would come to casa de Jorge to visit with me that evening. I wanted to be refreshed and wide awake when he came and hoped that Raul would show up to act as a translator. Later, my visit with Jose turned out to be more than I bargained for. He had brought with him a fellow researcher, Dr. Jacques Mabit. My conversations with them were quite useful (I'll tell you about them in later chapters), especially about using ayahuasca for inducing states of consciousness that enabled one to see into time. I didn't suspect then that it also would affect one's dreams.

After Jacques and Jose said good-night, left some reading material with me, and returned to their home, I felt too tired to read, and yet exhilarated at the same time. I went to bed, carefully fixing the fan so that it blew my way and fastening the mosquito netting around the mattress. I awoke in the middle of the night. I had two dreams.

The first dream brought Nonoy to me. We were sitting together and talking. I asked her, "How are you feeling about me?" She said, "I am afraid to love you." The second dream brought me in conversation with my mother, who died more than ten years ago. What were these dreams telling me? I didn't know. I soon fell back to sleep.

When I awoke, it was Monday, nearly midmorning. Surprised that I had slept so long, I quickly packed my bags. Jorge had arranged for me to ride on the back of his motorcycle, piloted by Ignacio, a neighbor who also looks after Jorge's home when Jorge is away. My plane was due to leave for Iquitos, where I was to join Jorge and his maestro shaman, Don Solon, at 1 P.M., and I had to be at the landing strip by 12.

When I arrived at the airport, I found, as was becoming usual, that the plane was late and was delayed until 3. At 3:30 the plane arrived and I got on it. However, the plane did not move. All the passengers were aboard, but no attempt was made to close the cabin door. The stewardesses remained at the front but nothing was happening. I felt as if I were on some weird ghost ship headed nowhere.

In a way I was right. After one half an hour, I finally got up out of my seat and simply asked a stewardess, "*¿Que passe?*" What is happening? She told me that the weather in Iquitos was awful. A huge wind was blowing and it was storming. They were waiting for it to clear before they would attempt to fly into it. I returned to my seat.

A few minutes later, I got up again. This time the message was *cancelado*, canceled. I wouldn't be able to leave Tarapoto until Wednesday, the day I had originally planned to leave.[2] It seemed as if that wind brought more than a storm with it; it also brought destiny.

For some reason, as I made my way out of the airport and found a taxi, I felt a sense of relief. Why? I immediately thought of Nonoy, the girl of

[2]I had made my reservations in Santa Fe, months before. I did not know then just how much time I would need to spend in Tarapoto and in Iquitos, so I had arranged my schedule to leave on Wednesday.

my dream the night before. What did this all mean? I went to Nonoy's home, quite happy to see her, and she was also pleased to see me. She told me that she wanted to show me more of the Tarapoto area during the next day, and she promised to come to the airport on Wednesday when I left to see me off.

She also wanted me to come to see her in a rehearsal for a play that evening. It was the play *Blood Wedding* by Garcia Lorca. I remembered seeing a film version of it by the Spanish director Carlos Saura just a few months before. I wondered what part Nonoy played. She told me that she was playing the part of the *novia*, an engaged young woman, who became strongly attracted to another man. More Spanish fire.

We arrived at an open-air auditorium—a schoolyard during the day—where the actors, dressed in street clothing, were assembling. Nonoy took great pains to introduce me to several of the actors. In my broken Spanish I explained who I was and what I was doing there.

Then Nonoy introduced me to her leading man. I immediately felt a sense of jealousy. Was I jealous? Or was he jealous? I couldn't tell. He seemed quite affable, and I noticed that he was in his thirties, a good ten or more years older than Nonoy. For some reason, up to that moment, I hadn't noticed how much older I was than Nonoy. She was, perhaps, barely twenty. I was fifty-four.

As if a gong had suddenly gone off, the male lead said to me laughing, "Who are you, her grandfather?" I was upset at this remark. I never even thought of myself as a father image to Nonoy, but a grandfather was just too much to bear. Not being able to use my mouth to retort, as I am wont to do in such situations, since my Spanish did not have the cutting edge honed as sharply as my English, I looked at him frowningly and silently. I was appearing laconic, even though that is certainly not my style at all.

I wondered then just how Nonoy saw me. Was I a father figure for her? Were my dreams of romance with her just silly illusions? In that sudden moment a truth that I was not facing in myself was revealed to me. I was not being truthful with myself. I was not seeing myself as others saw me. I suddenly felt old and a little sad. I also knew that it was my own illusions of myself that were dirtying the mirror of self-reflection. In my unconscious I heard an unspoken voice, "Clean up your act, Fred."

I watched Nonoy perform her part. I laughed when I realized that she and the detested leading man were supposed to marry, but she actually wanted another man. Soon the rehearsal came to an end. I walked Nonoy to her street, and before I took a taxi to Jorge's house, we made arrangements to meet the next morning.

The next morning, Tuesday, it was quite hot as I awoke. I seemed to have spent a dreamless night. I got out of bed, rolled up the mosquito netting, and, as the water was not running, dunked myself with buckets of cold water for my morning shower. I dressed and waited for Nonoy, who came by at around 10:30. After breakfast we went to the campus to tell Raul Espiritos that I hadn't left for Iquitos. We found him with an older professor, sitting at the palm-leaf hut in the center of the campus that served as a luncheonette and cold-drink counter.

We joined them at their table and ordered two lukewarm Cokes. It was growing progressively hotter, and I was becoming more uncomfortable. Raul immediately noticed Nonoy. He looked at me leeringly. Once I introduced them, he became very curious about her, asking her how she met me, what she did, and other questions. She told him about our participation with ayahuasca the previous Friday night.

He turned to me and spoke in English. For some reason he began to talk to me about love and how to make love with Peruvian women. I looked over nervously at Nonoy to see if she understood what he was saying, and asked her if she understood him. She said she didn't but she got the gist of the conversation. I wondered. Raul said, "Look, my friend, why don't you take her to the river and have a swim? Remember, with Peruvian women you must be aggressive, you must show her that you are crazy about her."

The older professor then joined in. He told us that although he is still interested in young women, they are no longer interested in him. He looked sad. I was a little embarrassed by the turn of the talk.

Nevertheless I felt emboldened by Raul's remark and decided to return to Jorge's home with Nonoy and see just what would happen. We left and walked in the hot sun and, after fifteen or more minutes, arrived at Jorge's home again. I took Nonoy inside and we sat down. I turned on the fan and closed the front door, wanting to be alone with her. We talked about my books. I showed her a copy of *Taking the Quantum Leap.* She was very curious about one quote taken from John A. Wheeler, "There is no law except the law that there is no law." We translated it into Spanish: *No hay una ley excepto la ley no hay una ley.* Nonoy was delighted with the paradox of this, and she laughed.

I was struck by her innocence and her charm. I sat by her, putting my arm around her, halfheartedly, not really knowing why I was doing it. She looked uncomfortable and I moved away. I then realized that just by that simple act, I had betrayed something between us. I had acted the wrong way. Our conversation seemed to come to an end. It was getting hotter. Nonoy said that she had to return home, and I suddenly felt very weary. She gave me a small hug and a kiss on the cheek and left.

I realized, dejectedly, that I really didn't want to seduce Nonoy, and I felt foolish that I had even thought of it. I sat down and felt hotter. I adjusted the electric fan so that it would blow full force on me. I got up and took another cold bucket shower (the water was still not running), and began to look through some of the materials that Jacques Mabit had given me. I always felt comfort from studying intellectual things. I needed comfort at that moment. I began to read.

A SEXUAL DREAM UNDER AYAHUASCA

Luis Luna[3] had collected many stories dealing with sexual energy and ayahuasca. Sexual abstention is more important than some realize. Failure to practice it often leads to misleading visions. Luna reported that one young man told his shaman, Don Emilio, that he had sex with his wife before taking ayahuasca. That evening he took ayahuasca. In his vision he saw a beautiful world filled with luscious and sexy women. They invited him to follow them. He couldn't resist. It turned out that he saw this vision while he was awake in a kind of sleepwalking trance. He left the ceremony.

The others thought that he had simply returned to his home. His wife, noticing that her husband had not returned after the ceremony, thought that he had remained there. The next morning they all realized that he was missing. They searched through the jungle surrounding the ceremonial region, but to no avail. They finally gave up the search. In the end they buried his clothes, as is customary in these cases. Several more days passed.

One morning a hunter noticed that there was something up in a tree. He was about to shoot it when he realized that the animal he was aiming at was a human being. He reported it to the police. An expedition was organized at once. They found the missing man in the tree almost naked, full of wounds, covered by worm and insect bites.

They took him to his home, where several women washed and healed his wounds. When he recovered his senses he was able to tell what had happened to him.

DUSK IN A DIFFERENT WORLD

I put my book down. I realized that I had been through a lot in the past few days. Everything was happening too quickly. My mind was spinning. Perhaps the shamans had been watching over me in Tarapoto. I wasn't to get involved sexually with Nonoy because the shamans knew what that

[3]See bibliography.

energy would do to me. Yet, I wondered, why had they put her next to me during the ceremony, and why was I offered the sexual potency potion by Wilfredo? I did drink it not knowing what was to happen to me that night. Was I being tested? Or was there some other reason that I took the potion? I had an uncanny feeling about the sequence of events that I had just been through.

Suddenly the lights went out. The electricity had stopped and I was plunged into darkness. I took out my small flashlight and searched for candles. Finding some, I lit them. The house seemed more peaceful under candlelight. I felt cooler and decided to walk out into the road and look at the stars. The now familiar Southern Cross shined down on me.

And again I remembered that I was really in a different world. Even the weather patterns circulate in the opposite direction down here. Water flowing down a basin circulates oppositely as it funnels down the drain. I am more than 65 percent water. Perhaps my cellular water was also undergoing a different circulation and this was churning up all kinds of suppressed feelings. I marveled under the stars, walked into the house, and soon retired for the evening.

I awoke early the next morning. Ignacio again agreed to take me to the airport. We arrived at around noon. Again the plane was delayed. It was Wednesday and I was sure that Jorge would be waiting for me in Iquitos with Don Solon, the man I had come so far to be with. I expected to probably take ayahuasca that evening, and I wanted to make sure that I didn't eat too much during the day.

After we arrived at the airport, Ignacio stood in line for me as I sat waiting for the plane to arrive. I looked around for Nonoy. But it was no, no, Nonoy. She wasn't coming to the airport to say good-bye. I never saw her again.

In the film, Nexy and Miguel had finished the ceremony. Nexy was to go off on her own, to search for herself and find her way in the world. Miguel would never see her again. And now I began to realize what Nonoy had represented to me. The ayahuasca had brought out a spiritual force in me. It had aroused me from my own self-pity, which was just another disguised form for self-hate. The jungle shamans were looking out after my soul. They carefully put Nonoy next to me during the ceremony. What I had felt was not really desire for Nonoy. I had felt my own feminine spirit, and Nonoy was the physical manifestation of that spirit. She encapsulated all that I had learned about the two sexual forces: The male force within me was the father spirit, the sky, the force of the conqueror, the explorer. The female force inside of me was the earth mother, the sacred, the inner space, the depth of my own unconscious. And through the ayahuasca, through Nonoy, through the *icaro*—the song of the healing shaman, I had heard both voices speak to me. I had regained, for a brief moment at least, an enchanted view of the universe.

I now knew where the magic in me was. It was in my feminine spirit. It was in my feeling-intuition functions, not in my thinking-sensation functions. Nonoy, as a spirit of the ayahuasca, had taught me an important lesson. I needed to "clean up my act."

The feminine spirits of the ayahuasca had also given me an inkling about what it felt like to live in the shamanic world. I had realized this before at brief moments.

I thought of a number of minor incidents in which that had happened to me. It had even happened to me in the middle of Washington Square in New York City. I became aware of the consciousness of the people around me and of the direction of their thoughts and was sensitive to a potential interaction with people that I was not desiring. The field of consciousness in that area was very powerful. This was primarily because there were a lot of drunks in the park and drunks are in an altered state of consciousness, making it often easy to pick up on their thoughts.

Any type of altered consciousness state can be tuned to. While most of us are in normal survival waking consciousness—the kind that says to us, "If I don't do this today I'm not going to get it done and I won't get enough money and I won't survive and I will die"—we fail to recognize that the altered state is still present.

In an active meditation or in an act of spiritual surrender it becomes easy to enter an altered state. Just spend a few days in nature with animals, and if you don't talk to humans for a couple of days, you will enter into animal consciousness very easily. You can sense trees, plants, anything. This has happened to me and I'm sure it has happened to most people. It is the most natural state of consciousness to be in.

The feminine spirits had also answered another question for me. They had given me a chance to see how the new physics relates in an artistic, poetic, and intuitive way to the enchanted world view of the shaman.

From a quantum physical point of view, there are two complementary ways of "seeing." The first way involves our intellect, our thinking and sensation functions. The second involves our heart, our feelings and our intuitions. The first way is masculine. It is primarily concerned with survival. The second way is feminine. It is primarily concerned with our spirit.

Sexual energy awakens a vibration between these two ways of "seeing." The sound of the jungle chiming, the crickets that had tuned to the resonance of the Tibetan chime, the feelings of attraction for the actress,

Nonoy, all were present to reawaken me to the unbroken wholeness of the universe. When I had felt this, I knew where my potency existed. It was in my reawakened feminine side.

I had learned an important lesson about healing and about the relationship between sexuality, magic, and potency. The reason a man falls in love with a woman is to regain the feeling-intuitive function within himself. He sees in her a mirror of his own spiritual self. The shamanic spirit had reawakened in my own sleeping beauty.

But what did this experience have to do with my understanding of shamanic physics? Gradually I began to see an answer. Shamans deal with the spirit world. Spirits play with us according to the shamanic view. We are prone to spirit infestation because of the battleground that goes on inside of us. It is the battle between spirit and matter. It is also the battle between matter and light. A person who is unbalanced in sexual energy between the male and female sides we all possess is also more prone to spiritual invasion than others.

Jung referred to the division between the masculine and feminine sides of a person's nature as the split between the persona and the shadow. The shadow was always the opposite sex of the person. When a man is in denial of his feminine shadow (his anima), he is in serious trouble. And similarly for a woman in denial of her masculine shadow (her animus).

I began to see that denial of the shadow was connected to the choice of the observer in a process. I pictured the observer as a sphere. The core of the sphere contained the shadow as part of the unconscious mind. The surface of the sphere was the ego. A man, as I saw it, would have a male ego or surface and a female core. I realized that this was a simplification, far too simple to explain Jungian psychology, but a physicist has to begin with a simple model.

If a man observes the world through his male surface, only without any information or interaction with his feminine core, that female core will begin to erupt. This eruption is caused, according to what I understood from my shamanic sense, by feminine spirits coming to the aid of the female inside of the man. It's as if that feminine core-spirit calls out for help from her sisters in the spirit world. She cries, "Let's wake this guy up."

A reversal would exist in a woman. When she denies her masculine shadow, her trapped core male spirit calls for help from his spiritual brothers.

Now this may sound bizarre. But I remembered, according to Buddhist thought, each of us is a composite of many entities. These entities vote and decide what we do each time we make a choice.

But how does one learn to pay attention to the shadow core that lies inside of each of us? The answer was simple. By paying attention to the opposite sex. For a man, a woman is an awakening of himself to his feminine nature. For a woman, a man awakens in her, her own "prince charming." The key insight was that we are mirrors for each other. When I am angry at some behavioral pattern I see in my feminine partner, I am really angry with myself, my own female side.

And now I knew where the healing comes in. When a person begins to observe with the whole sphere, including the core, a balance between light and matter takes place. Inclusion and exclusion are balanced. The body is in harmony with light and matter.

Nonoy had awakened me to my feminine shamanic spirit. That spirit began to speak to me in my dreams. She (I) told me that in order to become a healer, to really grasp what healing was all about, I had to come to a greater awareness of my own female core.

But I was still worried about other unwanted spirits coming into me. It wasn't until this moment that I realized that spirits can inhabit human bodies. Nonoy had a benevolent jungle spirit in her. But what would happen if those spirits were not friendly? I soon found out.

7

Shamanic Healing

Miguel is alone. Despondent, he wanders the dockside of the city of Iquitos looking for Nexy, and returns to the home of Meliton. He longs for Nexy and decides to continue searching for her. Wherever he goes he confronts the people of Iquitos, but they seem to ignore him. They even appear angry with him. He realizes that the people of Iquitos have become, for him, the spirits of the jungle. While in the physical realm, the realm of chronos, they appear to him as the people of the city of Iquitos; in the realm of mythos they are the spirits.

Later that evening, back in my hostel in Lima, my last day in Peru, I began to wonder about shamanic spirits. Where did they fit into the picture? And what about the realm of mythos? Perhaps myth played a role in shamanic healing. I slowed down and began to summarize my thoughts. I now knew that my first hypothesis was correct. Shamans do see everything they do in terms of vibrations. In terms of healing, these vibrations were sensed as being out of phase or of a different frequency from the healthy vibrations of normal organs in a sick patient. And as I had discovered, love and sexual energy were manifestations of these vibrations and, when felt, acted as healing energies, thus confirming my eighth hypothesis.

I certainly knew that my fourth hypothesis was correct: Shamans used any device to convince patients to change their beliefs about reality, in order to get their vibrations in harmony with those of the shaman so that healing could take place. I had seen such devices used in South Dakota with the Rosebud Sioux shaman, Doug White. But why was it necessary to suspend a patient's belief about what we perceive as reality? One would think that we needed to become more aware of what was real and learn to dismiss what was imaginary. Clearly this wasn't the case at all in the shamanic world.

I had also become aware of the shaman's higher powers and had discovered that they were connected with the planet Earth. Shamans seemed to gain power by being in their home territories. Even their potions and chants seemed to be more powerful when they were on their own turf. Although this fit with my seventh hypothesis, I wasn't sure why this was so. I speculated earlier about how radiation could have affected shamans. Perhaps it was a question of vibration again. Shamans had in their genetic structure the very ground from which they were born. It didn't seem logical, but DNA and RNA from Peru must be somewhat different from the same molecules found in Great Britain. I was guessing and feeling a little frustrated.

And I knew that shamans were able to construct reality to suit their needs, thus confirming my fifth hypothesis. They are able to see events as universally connected, and from those connections they are able to choose what they believe is physically meaningful, especially for an ill patient, again to match up vibrations.

But I was still unsure as to how every event could be connected to every other event. Somehow shamans were able to connect events that a patient, needing healing, was unable to. How did shamans see these connections? I pictured the "web of wyrd" as described by the shamans from Brighton. I saw God weaving that web. Perhaps they were seeing the secret of universal creation—a sense of how the universe constructs itself. Or, if you would prefer, how God creates the universe.

According to quantum physics, God plays dice with the universe. But if the cosmic game involves probabilities, then where are those probabilities evaluated? In God's mind? Probability is purely a mental concept. It is the stuff of dreams and myths. The die lies on the floor with six dots showing. It is only in my mind that I can assign to that position a probability of one sixth. It is only in my mind that I can "see" the die with another one of five possible faces showing. It is in my mind that probability exists.

Therefore, between any two events, who says what is likely and what is not? I flip a coin, I look at the sunset, and I watch the coin land. Who says that the events of flipping the coin and watching it land are more significant than watching the sunset and watching the coin land? I do. I see that there is a connection between the two events of flipping the coin and watching it land. I see that the two events are meaningfully connected while the sunset and coin-landing are not. I assign the probabilities by making the connection between the events. My mind creates the path connecting the events.

I then realized that consciousness itself acts in the universe by creating paths. And that objects in the universe do not follow single paths, but move as waves until the objects are observed. I remembered Werner Heisenberg, the founder of the uncertainty principle of quantum physics, saying, "The path comes into existence only when you see it."

Between two events there are many paths, many connections. By focusing on one of the paths, by restricting one's awareness to a single path as the most important, the other paths appear to vanish, even though they are still present. But what is so special about the path that is observed? It turned out to be a *least-action path*—one that appeared to bring order and sense into the world.

Then I realized that the habits we all have arise through the creation of *least-action* paths. Least-action paths were the routes we observed in everyday life. They were also literally pathways for neural information in our brains and nervous systems. By becoming conscious of the world around us, we created these pathways. They became habits necessary for our survival. But what about the other paths? Could we create new neural pathways by changing our thinking?

Between a causative event and an effect, all possible paths must emerge, according to quantum physics. Each path has an action in the universe. Some of these paths are quite remote and require more action, while other paths take less action. In many circumstances, particularly those involving the movement of large massive bodies, there is only one least-action path. It follows the laws of classical physics and thus appears to us as the order we perceive in the universe and in our lives.

But when we are dealing with atomic and subatomic matter or, as I speculated, with neuronal impulses resulting in thought and behavior, there are many least-action paths present, because all of the possible paths take the same action. In that case, the object moving along takes all of these paths simultaneously, if unobserved. We say the unobserved object is behaving like a wave. It occupies many paths simultaneously. But then, as a result of observation, one of these paths emerges as the actual path of the object. It becomes the least-action path. In terms of human behavior, this least-action path becomes what we've just noted as a groove or habit.

Thus at the quantum level, the action of a path can change, depending on the observer of that path. Each observation creates a least-action connection with the previous observation. By choosing to observe reality along a particular path, what was unobserved then becomes a greater-action path even if it would have been a least-action path if it had been observed. In other words, by choosing to observe a particular path from others, a least action is created. Consciousness creates, from all of the paths, the one that takes the least action.

I then rekindled an image I had just a week before when I was returning from Machu Picchu.

I had just visited the ruins of that ancient city, a short train ride from Cuzco, high in the Peruvian Andes. I remembered a young boy and the lesson he taught me. The group of tourists I was with and I had made our way down the steps of Machu Picchu to a bus waiting to take us to the train which would return us to Cuzco. As we got on the bus, this small boy dressed in Incan garb came up to us and looked at each of us as we boarded. He appeared quite serious and said nothing as we got on.

As soon as we were all on the bus, I looked out my window to see if I could see him. He was not in sight. But just as the bus made its way down the first hairpin turn leading to the train station, I saw him. This time he was smiling at us and waving. And then just as we left him, he let out a sound that was piercing and funny at the same time. It was a "hoot." It sounded like a loud owl cry. The sound pierced my whole body and made me shudder.

He then vanished from sight and I thought no more about him. The bus was rounding the curve to the second lower turn in the descending road when another boy dressed similarly to the first suddenly jumped out from the bush bordering the road and made another "hoot." He looked surprisingly like the first one.

With each turn down the sloping path another boy jumped out, and each time I felt the same thing. I thought that there were several boys dressed alike, all of them stationed along the twisting road. I believed that each of them was part of some tourist show. But I was wrong.

Soon we were at the train station. When we got out of the bus, the boy was there waiting for us. Somehow he had transported himself down the mountain ahead of us. I then realized that the "boys" I had seen at each turn of the road were just the one boy.

He had been enacting an ancient tradition of the Incas, performing a ceremony known as the *Run of the Chasqui.* These runners carried messages down the mountain from the saddle-high[1] city to the workers below. They also were tricksters and reminded me of the ancient god Hermes or Mercury, the messenger of the gods, in their antics. They were carriers of the truth, and even though they acted up, they could be relied upon.

[1]The city is situated in a saddle between mountains with dropoffs to the sides.

The boy had been racing down the mountain, taking a new least-action, straightest pathway, and thus was meeting us at each turn of the road. The bus was taking its normal "survival" course down the well-trodden habitual path. This path was an analogy of our own habitual thinking. To us it was the safest pathway down, and since we had taken it, it was, for us, the least-action path. Our action was governed by the knowledge that any other way was forbidden to us, if we wanted to survive.

The mythic path was there, but we hadn't realized it, because it wasn't safe to do so. Of course, any of us could have left the bus and followed the runner down his exhilarating path, if we knew that it existed. But none of us did this. None of us was aware that any other way was possible. None of us, at the level of metaphor, was aware of a mythic reality.

In my imagination, the runner was carrying a message of truth from a different time, a mythic time. He was a reminder that I need not take life so seriously. If I could tune to a richer mythic sense of myself, which had become more and more real with each passing moment of my life in Peru, I, too, could find life to be more enjoyable, healthier and fulfilling.

There is a fine line between truth and myth. Often writers must cross that line in order to seek information that will bring strength to whatever they are writing. I was realizing that, as a quantum physicist and writer, I had skirted that boundary and then darted across the line back and forth many, many times. Again, I began to wonder just what the truth really was.

Quantum physics forces you to do this. Its very nature deals with an invisible world, and consequently all we physicists can do is invent myths we call models to explain how that invisible world behaves. Out of the behavior of invisible matter and energy comes the behavior of visible objective matter and energy.

No one really knows why quantum physical laws work the way that they do. We do know that matter on a subatomic and atomic scale does not behave in any way like large objects do. In fact, physicists must deal with such matter not so much in terms of what matter is and even how it behaves, or what laws of movement it follows, but in terms of all the possible ways in which it can change from one state to the next. It is not enough to limit oneself to a fixed path. Things will go from A to B. But they follow every possible way to do that. As I realized after the bus ride, somehow there was another least-action path that none of us had discovered yet.

And I was looking for another law of the universe, one that seemed to show how to choose one form of reality rather than another. If I could find it, then I would know how the shaman seems to be able to break the law of habitual observation by choosing nonordinary reality.

Nonordinary reality? Was that mythic reality? Was mythic reality a new least-action pathway in the universe? Was that the spirit-reality of the shamans?

I recalled my second hypothesis. Shamans were able to see the world in terms of myths and visions that at first seemed contrary to the laws of physics. Perhaps there was a connection between how shamans choose reality and these myths and visions. Then I remembered the movie spirits.

Why did the spirits ignore Miguel? Was it because he had insulted Nexy in some way? Perhaps they were there to nudge him along a least-action pathway that he hadn't even considered before. What was missing in his life? Dimly, at first, he realized that he had failed to come to grips with his feminine soul. He, by seeking for her outside of himself, had failed to realize the message: She lives inside of him.

As I remembered Miguel's search for Nexy and his confrontation with the people/spirits of Iquitos, I remembered the day I left Tarapoto and my own meeting with the angry spirits of the jungle. They taught me that spirits were real, and they nudged me into a new least-action path in my own life—an awareness of mythic reality.

I was waiting at the airport in Tarapoto. The plane had still not arrived. But the loudspeaker said that it would be there in fifteen minutes. It was nearly 3:30 P.M. I got up and took my place at the front of the line where Ignacio was holding his position against the onslaught of pushing and worrying passengers. When I was established in line, Ignacio wished me well and said that he had to return to his home and business. I thanked him and gave him a small gift, for him and his family. He was very grateful, thanked me, and waved as he left.

Finally the plane landed and the crowd behind me grew progressively worse. A group of three people, consisting of an extremely fat woman and her two male friends, all smelling of foul body odor and alcohol, were just behind me. Feeling both friendly and hostile at the same time, they kept pushing on me, reducing my standing space to just a few inches. I asked them to desist. They then decided to have some sport with me, calling me *tio* (uncle), a term used derogatorily to mean an old man. I remembered the actor in Nonoy's play who thought I was Nonoy's grandfather. I felt very old and was upset. I said to them, "*¡Me no tio!*" rather angrily. One of the men became abusive, shouting out swear words in English. "Fuck, shit" was all he seemed to know. He somehow knew I knew what they meant even though I had spoken only Spanish, but he felt safe to say those words there since probably few of the people in line spoke or knew any English.

I stood there in silence. Again my mind was at work. Why was I being abused this way? I thought of Nonoy. I began to feel again that I was somehow on a journey that was more than just a search for truth. It was a search inside of me, a journey into my mythic realms of being. With all of the pushing, bad odors, and abusive language, I felt a mythic level suddenly present.

You'll recall that I saw that Nonoy was a spirit of the ayahuasca—a spirit from a time long ago. She was a reminder to me to find my feminine soul inside of me. She had been there to connect me with my own intuitive-feeling self. This "inner" self did not operate in Newtonian clockwork time as my physicist's mind did. Nonoy had brought me to a new vision of time, a mythic time, which does not run alongside clock time. Instead it runs in another direction, perhaps perpendicular to clock time. I did not know. I was only feeling grateful that I had experienced it.

Suddenly all of the images—Jorge's sexual pots, Nonoy, the ceremonies I had been going through, and even flashes of my life—came into vision. I realized that with ayahuasca both the future and the past can be "seen." Rather I would say that I "saw" all of this in a kind of timeless state. It was as if there was no time and that the experience I was having at that moment was beyond physical time. I was having another mythic time experience.

Again a jolt from the rear. "Fuck, shit" brought me back to the physical time. But the mythic time lingered. These were angry spirits, not people behind me. How dare I attempt to seduce Nonoy? She was the spirit of ayahuasca, not a sex object for me. I then realized what was happening. I turned around and smiled at the three rude gods behind me. Suddenly they stopped pushing and smiled back at me. My rude gods had become friendly and respectful.

Then, suddenly, we were all pushing forward to get on the plane. This time the plane was going to Iquitos to meet Jorge again, and more important, I was going to meet Don Solon, the shaman who had taught Jorge.

I watched Miguel continue his search for Nexy. While he gradually realizes that she is no longer part of his life, he refuses to give up. He goes from place to place, retracing his path, seeking out the old places where Nexy had spent time. He sees that Nexy was a part of himself and that she had opened him to a new vision of what he was to do with the rest of his life. From this point onward, Miguel had to practice a new way of living. He must tune into his own feminine intuitive-feeling self. But this was not going to be easy for him.

And it wasn't going to be easy for me. I was after the secrets of shamanic healing. I thought that I could get them using my rational masculine self.

In hoping to do this, I had divided my own consciousness. In terms of the Nobel Prize-winning physiologist Roger Sperry's split-brain model, part of me was the left-brained rational side and the other part was my right-brained intuitive side. I had hoped to realize the secrets of shamanic healing using only half of my consciousness: the left.

But I now suspected that this would not work. I was learning far more than I had expected. I now was beginning to suspect that the real secret of shamanic healing was not to be found using only one half of my own consciousness. Somewhere inside of me, all of these experiences I was having were pointing in only one direction. It was said a long time ago by an ancient seer, "Physician, heal thyself."

I had to integrate the two parts of my own mind. I had to bring them together. Even more, I had to heal my feminine self so that it would contribute equally with my masculine self. How was I to do this?

I saw an answer. See your own truth in the world and tell it the way it is. All of the old shibboleths were coming back to me. "Tell the truth." "To thine own self be true." "Love yourself, before you love another." "From compassion comes real power."

But I felt that I had lost that power. What is love? Even though I had explained love as a form of vibrational energy, I began to think that I did not know. All I really knew, when I looked inside myself, was I knew fear. I was aware that I was afraid most of the time, if I really got in touch with it.

But how was I to regain my own sense of myself, my own power? How was I to really heal myself? Perhaps the master shaman, Don Solon, would be able to do for me what the movie shaman was doing for Miguel.

When I had first met Jorge and his translator, Mike, in the United States, Jorge said, "If you think I am powerful, my maestro, Don Solon, is ten times more powerful than me." Mike, having met Don Solon, had added, "Iquitos is a big port on the Amazon. Don Solon is a seventy-year-old wizard that lives in Iquitos. He is an herbal *curandero*[2] and he is a powerhouse. He is outrageous. He is like a wall and he can direct visions like no other."

LEAST-ACTION PATHS AND THE UNCONSCIOUS MIND

My plane was touching down at the Iquitos airport. As the plane was landing, I realized that it had left the ground at Tarapoto a little after four.

[2]Remember that a *curandero* is a healer, one that cures illness.

We were up in the air for only ten minutes when the pilot announced that we were landing in Iquitos. By air, the trip was less than twenty-five minutes from the close of the cabin door at departure to the open cabin door at arrival. Yet to attempt to make the same journey by land would take several months, since there are no roads connecting it to any other city.

Iquitos is the capital of the department of Loreto and the chief town of Peru's jungle region. Indeed it is the largest jungle city in the entire Amazon basin which is not connected by outside roads. It is situated on the west bank of the Amazon River some 3200 km (1980 miles) from the mouth of the Amazon. It probably has a population of 200,000, although it is reported at 350,000.

Iquitos has quite a varied history. Founded in 1750 by Jesuit missionaries, the city barely happened. It seems the local Indian population didn't like the white folk coming down the river and fought them tooth and nail. But the city survived its hostile birth and grew slowly until 1870, when the population reached around 1500 (not including the Indians, of course).

The city had seen its day. It had passed through a rubber boom and then a bust. However, a new hopeful possibility had recently arrived in this jungle city: harvesting medicinal herbs and plants. A team of young biologists, herbalists, and horticulturists, coupled with archeologists from some eastern universities in the United States, had arrived in Iquitos. Their mission was to help local Indians harvest the many varieties of plants and medicinal herbs in the jungle.

There are more undiscovered plants in the Iquitos area than there are identifiable plants to date, according to many who are presently researching the area. By teaching the local population how to harvest and sell what they harvest, these young "Schweitzers" hope to make the Indian population what they were before the white tide took over.

As I walked from the plane into the airport I was greeted by Jorge. He hurried me into a taxi, and as we climbed in, he introduced me to his companion, Arturo, who was a tour guide that Jorge had met the previous Saturday. He told me that Arturo had just started a jungle tour company and had agreed to take us all an hour's boat ride down the Amazon River to a jungle retreat. Arturo was instructing the taxi driver to rush to Don Solon's house to pick him up. There was to be a ceremony that night, and the boat was waiting to take Don Solon and us to the retreat. Don Solon's house was on a small unpaved street in an industrial neighborhood. As soon as I entered the open doorway, I was greeted by his young wife and two of his daughters. They were quite delicate and very charming.

Soon Don Solon appeared. I shook his hand and told him that I had come a very long way to meet him. He was a tiny man by our standards, standing about five feet four inches, quite thin with a small black moustache, balding forehead, and large ears. His arms, which appeared quite strong, were covered with thick black hairs. Dressed in the appropriate clothing, he could have appeared to be a good image of Captain Hook in the Walt Disney animated version of *Peter Pan*. His features were more Spanish than Indian. I guessed that he was a mestizo. He carried himself with a proud bearing. Aside from the graying of his long sideburns, his hair was black. Don Solon was seventy years old, but as Jorge told me later, "He is old, but he is strong."

I noticed that Don Solon's right eye was nearly blind. The iris was covered with a blue cataract. His left eye, the good one, was dark brown. Later he told me he lost the sight in his right eye due to the spell of a *brujo* (an evil magician).

I had brought with me a number of gifts, and after I had given them to Don Solon and his family, Arturo came in and said we must be on our way. The sun was just beginning to set as we boarded our motorboat and made our way down the Amazon River. Don Solon was sitting to my right. Jorge and Arturo were behind us. As we rode, the air passing across our bodies grew pleasantly cold, and I found myself wrapping up and closing my collar to keep warm. Don Solon had brought a jacket with him. Jorge leaned forward and said to me, "You have made Don Solon very happy."

Soon the sun had set and we rode along in complete darkness. I noticed several lights along the shoreline. These were from fires from small native villages. Even though we rode in a motorboat, I felt as if we were being guided by the flowing waters of the Amazon. My mind was playing with the idea of how rivers form with all of their meanderings. Why didn't they flow in straight lines? What restricted the rivers from finding the shortest least-action paths? Perhaps the straightest line wasn't a least-action path. But why not?

I then began to see why the Amazon and other rivers didn't flow in a straight line, even though they were following least-action paths. They weren't straight because, in a real sense, the earth was "observing" the flow. She had her say, too. She determined the way waters flowed on her surface by restricting the pathways.

I was feeling very excited and elated. This was the moment I had been waiting for. I had finally met Don Solon, the master shamanic healer. But I still was thinking about the physics of shamanic healing. As we glided through the dark, I felt as if I had really entered another world. And then, suddenly, during this moment of reflection, I began to think once again about how the universe created this world. It had something to do with how the river formed. But my thoughts were becoming hazy, as hazy as the dusk that was quickly descending.

Suddenly, Candace Lienhart, the American shamanka whom I had known for several years, popped into my mind. She had first told me about the physics of shamanic healing and about habitual paths in consciousness. My mind drifted back as the boat motored smoothly along the Amazon River.

In late September of 1988, I had returned to the United States from Europe. I wanted to speak with Candace Lienhart, whom I first met in 1983 in California, who was now living in Arizona. I called her and she agreed to meet me.

Candace Lienhart is an unusual woman. She is, in my opinion, a full-fledged American shamanka, if I am allowed to use that term to describe her. As in the shamanic tradition, she has had her own personal encounter with death and a shamanic teacher.

Candace, being an American, coming from a background similar to mine, was easy to talk to, and she knew just what I was doing and why I was doing it. I had spoken to her just before I flew to Europe. When I told her what I was about to embark on, she warned me and she told me that she would put up a protective "circle of light" around me as I traveled into the "new world."

I met with her in her home in Tucson, Arizona, after I returned from Europe. I first asked her about the connection between healing and habitual consciousness. She told me that most people don't realize that the universe is totally neutral about who we are and what we can accomplish. Many of us feel that the universe is malevolent or, at best, because of past conditioning, because we were treated badly as children, think that this is the only way we can be treated. So we continue to wish for bad treatment, unconsciously.

Anything that becomes habitual eventually becomes "unconscious." Once we become aware of a path of our human behavior, i.e., create a least-action pathway in our nervous system and brain, our mind begins to observe in a different manner. The habitual paths form what we call the unconscious mind. It is really a misnomer. The unconscious is really very conscious, and it is observing all of the time. In our mind we make a groove, a roadway, that we follow without thinking. People stay in the groove, even if they are uncomfortable. They don't even think about it anymore.

Now the question is, how do you break out of the least-action groove?

Candace explained that the groove is imagined as if it were in a phonograph record. The needle of life, our conscious awareness, is put in the groove. In order for the needle to not hit a particular groove, the groove must be filled. This is not a surgical procedure. You don't take a drug to do this. You actually don't take the groove away, you fill it. She asked me, "Do you understand this?"

I asked her, "Does filling it mean that you grant the desire of the groove in some sense? What does that mean exactly?"

She explained that filling the groove was like offering an equal and opposite vote to any action that was anticipated.

I knew what that meant. In order to vote, it would be necessary to become aware of the action that you were doing or going to do. Usually we are unaware of actions that we take because our beliefs are based on past experiences and those in turn are based on the habits we have built up. It is a self-consistent loop. And it is a trap.

Candace said, "For example, if you believe that you are fearful, then you will be fearful."

I realized that if you have built up that belief over many years, then it has become part of your body-mind. It has become a large groove. That meant that it was a groove that your consciousness "needle of life" would be strongly attracted to.

Candace explained how the needle in the groove worked. When the needle comes to a set of possible grooves, a choice is made and one groove is picked out. She said that it was a majority rule going on inside of you, just like the government of a country. That government acted in a particular way according to the voting records of its membership.

Candace's metaphor reminded me of a story that was attributed to the Buddha. It told of a village where a group of persons sat. The village consisted of senior citizens, teenagers, children, working-class people, rich people, poor ones, wise men and women, councilors, transients, and even tourists. It was their job to decide the future of the village.

They sat for long periods of time and discussed what to do. Now and again one of the members rose, gave a speech, and suggested a certain course of action to be taken. Sometimes, several members, particularly when there was a sense of competition between an elder member and a junior member or two or more junior members, rose and spoke at the same time, often in demanding tones, insisting that he or she was right and deserved the full attention of the council.

Even violent quarrels took place in this queer group; some of the members at times ganged up on others to injure them. Sometimes it got too tough and some people left the village. And as it would be, new members came to the village, liked it there, in spite of the quarrels, and decided to stay.

Since some members were quite old, after a while they died. Some were feeble and voiced some concern about "the way things were going, how we must all follow tradition," etc. Sometimes a few of the youths became inspired with the words of these elders, listened to them and decided to follow their ways. This emboldened the old ones and gave them strength to continue the battle. Righteously they mustered forces in the assembly and began to speak louder and more demandingly, insisting that now they were the majority, the "moral majority," and all should follow their dictates.

The Buddha used this village as a metaphor for the body-mind. This village is you. The members in it are your physical and mental elements. They are your instincts, tendencies, ideas, beliefs, desires, etc. Each has its own parents going back dimly into time. Each is, if you like, a *soul*, a person. And yet there is no one really there. Each calls himself, herself, ourself, a *self*. Each thinks that she is in charge, he is in charge, they are in charge, for a brief splitting nano-nanosecond of time. But, after a while, each *soul* sits down to the inevitable fleeting of the moment.

The main agenda for the voters is survival. And that means consideration of everything that we are afraid of—our fear being based on what threatens our survival. Whenever there is fear in the system, the agenda is not complete. The record groove is not filled. I realized what it would take to fill it. You had to move the needle through it. When you have moved through a fear, the groove fills, and the "needle of life" need not pass through that groove again. Candace said, "There is no magnetic draw anymore, if you have fulfilled all of the fears that you think are possible."

This meant to me that one changed habits by changing how you moved from a causative event to an effect. By not facing one's fears, by not moving through that particular groove, a habitual unconscious path of avoidance is set up. Then these unconscious paths appear habitual because they become least-action paths. These paths must be connected with survival.

Thus, I reasoned, the universe itself is concerned with its own survival. The universe we experience is the result of consciousness creating and then following least-action paths.

My logic was peculiar. I had an unconscious mind; therefore, the universe was operating unconsciously. But the universe was evolving. That meant to me, at least, that the universe was becoming more aware, more conscious of itself. And sentient life-forms were the organs through which the universe became aware. Once one became aware of the least-action paths, they would no longer be paths of avoidance. By becoming aware of these paths of avoidance, and risking taking the paths we are afraid of, more action would be created. New pathways would emerge. Then these other paths would become the least-action paths, and other habits would ensue.

I suspected that it would then be possible for a universe to follow other least-action paths, but without fear. But how did fear arise in the first place?

Candace told me that she had a spook[3] come to her at one time and say, "Candy, how much do you want to freak yourself out here?" Candace thought, "Do I have a choice?" She had thought that this was just happening beyond her control. The spook said, "No, it's your choice. You decide how much you want to freak out." Candace said, "It's enough already." And the spook agreed and then changed.

Candace added, "But in changing our minds, that's the thing that people get so stubborn about, they forget or they don't know that the body-mind has the majority rule over your conscious mind. The body-mind is holding that accumulative recorded message of all the 'should-bes' and 'what-ifs' and the 'maybes' and all that stuff in your system."

This began to make sense. Certainly our bodies, our cells, follow the laws of physics. To survive, certain habits of nature had to be present. Without them, the species would have become extinct. Thus our body-minds, through the passage of time, evolved appropriate least-action paths. If we were simple creatures like cockroaches, there would be little need to change those paths. But human beings evolved and, through that evolution, learned to alter which paths were least-action. The problem is that many people have not learned to adjust to new environments, new situations. They continue to rely on old pathways. They follow old habits, believing that these will ensure their survival. But it turns out that these paths will do no such thing. Instead they will ensure that the person paradoxically does not survive.

There were many examples I could think of. Take our use of tobacco. In the shamanic villages, tobacco was used as a healing medicine. Today, in our society, it acts as a killer. The least-action path of tobacco use fit in the village or tribal days. It doesn't fit into our society.

Take overeating. Our bodies have adopted the ability to take on fat when we overeat. It isn't necessary that our bodies be this way. We could eliminate the extra calories by simply raising our metabolic rate or simply eliminating them in our feces. But we don't; we get fat instead. The least-action path of survival produces body weight. It remembers the days when we hunted for food, and stuffing ourselves was our way to survive a long cold winter without game.

Our mental habits associated with such things as smoking or overeating are buried very deeply. Those grooves are wide indeed. If we had no fear

[3]She refers to spirit entities as "spooks."

of starving, we couldn't overeat. If we had recognized the spiritual use of tobacco, we wouldn't oversmoke.

I then realized that this was the crucial point I was seeking. Healing was a function of freeing ourselves from inappropriate least-action neural and physical pathways. The answer was simple to state but difficult to do in practice. Make conscious that which is unconscious. By so doing, new least-action pathways would emerge and a new state of health would occur. But now I needed to see how a change in least-action paths could act to heal someone. I asked Candace how she used her ability to free a person from nonbeneficial least-action paths.

She told me that she does it in several ways. Her first question to a patient is, "Are you willing to release this habit or way of thinking?" Usually a patient knowing the consequences of maintaining the habit will say yes. Now when the body hears the mind and the mouth say that they are willing to release the habit, it can either say yes or no. Sometimes Candace had known the body of a patient to say no and the mouth to say, "Yes, I am ready to be well."

Candace said that the body has been listening to every internal thought, and they all have been recorded. And each thought has been put in that majority rule–minority rule category. So the votes are being recorded by their thoughts continuously.

For a body to say no, the foundation that the illness was built on was very strong. The person wanted to be in a major drama for a reason, but the mouth couldn't admit that he really wanted it.

She said, "I ask their mind, 'Do they want it?' Then I ask the body, 'Is that possible?' Sometimes the body says no. Then I have to negotiate with the body-mind. If the body-mind says, 'No, I'm doing this and I refuse to change,' which it seldom has done in my experience, I can't do anything.

"One time a little guy,[4] inside a client, came up and said, 'No. I'm not going to do it.' So I asked, 'Why not?' The little guy said, 'Because this guy has been running this number on me for twenty-five years and he's been telling me that he really wants it. He wanted cancer. He wanted heart disease and he wanted something horrible. And that's what I'm making for him. And now he's telling me that he doesn't want it. I worked all this time to have the body be as he wanted it.' "

So, I realized, the body-mind has the power to create illness. I then told her that I would like to understand what shamans sense when they heal.

[4]The "little guy" was a spirit entity that Candace sensed when healing the patient. He was the voice of the body saying that he wanted to follow the instructions of the mind.

I suspected that there is a certain level in which the physical world, the world of matter, and the world that a shaman manipulates, the world of spirit, begin to overlap.

I thought that perhaps a healer saw a certain connection between these worlds. Chris and Richard, the Brighton shamans, had told me (see Chapter 3) that according to the *wyrd* way, spirit and matter were connected. But I had never seen a spirit nor any manifestation of one. I wondered if she saw, perhaps, an aura or a field surrounding the body. In watching Candace work in the past, I noticed that she seemed to be manipulating something without touching the body at all. I asked her, "How do you sense that a certain organ is going wrong? What is it that you are picking up on?"

NOTHING BUT VIBRATIONS

Her answer turned out to be universal amongst shamans. "What I look for is an imbalance or a disturbance."

"How do you sense an imbalance or a disturbance?"

"It's a vibrational experience for me."

"Something vibrates back and forth?"

She told me that organs in the body vibrate. She said, "Every organ runs at a different frequency, and every body runs at different rates of speed at different times. So if there is a disturbance in the body, it appears as a deeper density. It's thicker, it's more massive. It's an interference. It's a speed interference."

I was confused by her answer. I couldn't see how an interference was at the same time a density. It wasn't until much later that I saw it.

I said, "So it's an interference and it's a density. When you use the word density, I am not sure what you mean. Here is how I think of the concept of density. I have some objects in my hands. They have the same size. I pick one up and it weighs more than the other so I know it's more dense. Are you doing something equivalent to that?"

"Yes."

"Okay, could you explain what you do then?"

"I use my hand as a receptor to perceive. My brain is an observer. My hand is the receiver. As I move through the density of the body, any disturbance or any density that is dragging, I see as an interference."

It was only then, at that moment, riding in the boat to a jungle retreat up the Amazon River, that I knew what she meant. In my college physics days, I remembered doing experiments with light and photosensitive films. We would make light wave interference patterns. By allowing light to pass through a screen with two slits, for example, the waves of light coming from the slits would reach the film and make a pattern on the film. Because the light waves were vibrating, they would reach various points on the film in or out of phase with each other. We would observe a typical light wave interference pattern. To measure it, we had to use an instrument called a densitometer. It enabled one to measure how thick or dense the darkening of the film was in various locations on the developed film.

Candace was a shamanka who operated in that crack region between the two worlds of matter and spirit that Richard and Chris had told me about. She had told me earlier that when she healed someone she sensed interference patterns using her hand as a kind of antenna. I reasoned that these patterns could be overlaps of waves, perhaps similar to the interference patterns produced by electromagnetic waves. Such interference patterns were even present in atoms and molecules, only the waves that overlapped were quantum waves of probability.

These waves are the guiding waves of matter. Where these waves overlap positively (the crests of the waves match up), matter is more likely to manifest. Where they overlap negatively (the crests of the waves do not match up), matter is least likely to manifest. The interference of these waves was crucial to where matter would manifest.

I looked down at the waves as the boat made its way along the river. I remembered our meeting again.

Candace explained that there are two types of interference. One is created internally and one is an external interference. They are neither self-generated or allowed to enter the body. When she sweeps her attention across the energy of a body, she looks for differences. She said, "When people are in their fullness, their energy will hit the edge of their hands as they are extended out in the world. When a person is frightened, he or she will contract that. And the energy pulls into a little dot inside their system."

I realized that Candace was describing her experience dealing with subtle or other-world matter. Perhaps she was dealing with matter from the mythic or imaginal realm. I didn't think she was referring to the actual density of the body. She was perceiving another reality.

I asked her, "Is there a particular place where that dot or knot forms?"

She told me that wherever it forms, the person has contracted his consciousness into a knot. It's like a dent in the energy field. Where the body has been hit with something, there is a recoil. Where it has been damaged or it has been bruised, there is a little conflict going on at that spot.

Candace confirmed that this knot-manifestation started on a thought level first. She said it was magnetic.[5] It all begins with a question; perhaps a doubt or insecurity is enough to start the process. The consciousness of the individual says, "I wonder what will happen if . . . ?" And that one question of faith or fear will draw the contraction in.

The depth of the contraction depends on how much you invest in the thought. Your desire for manifestation is the attractor. It will attract from the cosmos, from the environment of the person; it will attract a degree of density that matches the person's desire. To the degree that the individual wants to investigate it, it will come right into the interior structure of the body, into the physiology.

So the desire begins in the person's perceptions or thoughts. The amount of energy that is fueling these perceptions and the speed at which they travel decide how dense they will be in the body. For example, you may say, "I want an ice cream cone." And it could be a fleeting thought. It's not a big desire and it doesn't strike you as something overwhelming, but if you say it twenty-five times, it will engage with or connect to what you desire.

Candace explained that with regard to these thoughts we are like radio receivers and broadcasters. We are both sending and receiving answers to our thoughts at the same time, all the time. What you want, actually comes magnetically to you by way of the degree of your want. That's what speeds it up or slows it down. And what you want can also be what you fear.

I began to see this action in terms of pathways between the present (lacking something) and the future (having that thing). According to the transactional view of quantum physics, there are an infinite number of pathways connecting them. Associated with each pathway is a probability of successful connection. If the present and the future events are in harmony with the universe, the probability of success is great and a least-action pathway will be present. Desire was simply the awareness that what one wanted and what one could have were in reality the same thing.

She pointed out that this action of receiving and sending could be simultaneous, but not necessarily. It was more of an action-reaction business. It was like having a menu and placing an order. Every thought that you have is like a pebble hitting still water. And the waves of that thought go out until they hit the shore of what it is that you are asking about, and then it starts waving back to you.

I realized that this ability to alter reality, by desire, thought, wish, or

[5]When Candace uses the word "magnetic," I don't think she means the same thing as what physicists would call magnetic, although the effect at the subtle level could indeed be the same as the physical effect of magnetism on matter.

whatever, is really nonjudgmental. The self[6] doesn't care if it is something that is going to harm you or help you. The universe is neutral in fulfilling your desires. I asked her if a person could get anything he or she wanted by putting the thought out there.

Candace said, "Oh, yeah. Anything you want. That's the universal law. It totally doesn't care. The universe has no judgment. If you say, 'I'm going to bring in the stuff that's going to kill me,' then universal law says this is the way the law works so this is what you are going to get. If you are going to bring in the stuff that's going to make you wiser or wealthier, it's going to do everything it can to provide you with that."

Back in the boat, motoring down the Amazon to my destiny with Don Solon, I realized that was it—a simple shamanic truth. Our bodies are reflections of our thoughts. Ordinary doctors do not deal with the connection between the body and the mind. Shamans do. They are able to "see" just where thoughts are embedded in the body. They do this by sensing interference patterns in the fields surrounding the body.

Candace was using her body as a kind of densitometer. But instead of detecting light wave film emulsions, she was detecting another form of vibrational energy. Again I was reminded of the same lesson that Jorge had taught me a few months ago when I met him in the United States. He had said, "Life is nothing but vibration." I turned around and looked at Jorge. He was smiling at me. Although I could barely see his face in the dim light, I had the uncanny feeling that he knew what I was thinking.

As the boat rocked along, the song "Row, row, row your boat" came into my mind. Was life nothing but vibration? Or was it something else?

NOTHING BUT A DREAM

It was becoming very apparent to me that mind had a far more serious role to play in healing than even I had ever suspected. I had a number of flash insights as we were moving along the dark bloodline of the Amazon. I knew then what I had been missing all along. Imagination. Without it, it was impossible for the body to heal. We all have built-in body images. That image is remembered in the auric field or morphogenetic field surrounding the body. It accounts for healing the body, because it remembers just where the molecules must go in order to heal the cut or mend the bone.

It was dawning on me that it just didn't stop there. We could help the process if we could use our minds to imagine the process of healing. In

[6]Or the conscious universe.

the imagination, there is no distinction between the mind and the body. The body responds to the mind, and the body, in turn, cultures the mind by providing physical boundaries which limit the mind to the extent that the mind believes it is the body.

But by extending our minds beyond our bodies, something that we can learn to do, we can actually alter the body to suit our dreams. Was I crazy? I didn't know. But in talking with all of the shamans I had met so far, each of them was able to envision the process of healing by using their power of imagination in a far more meaningful way than I was.

I then remembered how the Chumash doctors, Kote and A-lul'Koy Lotah, used their imagination when they healed. They too detected interference patterns.

I had asked them, "Do you see the healing process in terms of spirits or the spiritual realm? I realize that this is complex. What happens when somebody gets ill? Could you give me an example that you have dealt with. What did you do to heal that person? What was involved in the healing?"

They told me that they dealt with a tremendous number of women who have had abortions. Women will come to them and they will have a reproductive problem. Nothing is going right. Their monthly cycles will be messed up. Their whole rhythm has been violated. The two shamans then see a pattern.

Kote said, "I ask them, 'How many abortions have you had?' A typical woman would say, 'I haven't had any.' Then later on in the conversation, again I'll ask, 'Did you have at least one abortion? You must have, because this and this is happening.' She'll say, 'Well, I've had one, when I was this age or that.' "

Kote explained that when conception takes place, the body goes into a seven- to nine-month birthing mode. If that mode is interrupted, sometimes the body will continue following a course even though the process was physically stopped. The emotional continuation of that mode will go on. So now the body is in an emotional birthing mode, but it's also trying to do its monthly physical cycle. So the rhythm is broken. Also at the end of that nine-month time period, the chemistry in a woman's body changes, especially in the whole area where the child is to be released. The aura or energy field around the missing child may still be present.

CHUMASH X-RAY

I recognized that the Chumash medicine people, like Candace, see the auric field. I asked them to explain what they saw.

Kote said, "We use the aura as a diagnostic tool. It has nothing to do with spirituality. We call it Chumash x-ray."

"What is the aura and how do you see it?"

"The aura is a glow of energy around the body. It depends on the individual, the color it has, and everything. But if I'm looking at somebody, if I see some shadow there, it doesn't represent a hole; it means a chemistry imbalance. So if you're imbalanced and your whole body is glowing a kind of yellowish-blue color and all of the sudden there is a purplish color by your liver, I know that the chemistry has changed and that the heat exchange has changed. Now this doesn't tell me what has happened, but it tells me that something is out of balance. The liver deals on the emotional level with anger. Anger is a symptom of frustration. Frustration goes back to having a judgmental attitude. 'Who were you judging when you became angry?' "

"You see this aura, but most medical doctors can't. So how do you see it?"

"Doctors have too much doubt in them to see it."

"So what does one do to see the aura?"

"I want to. It's there."

"So the reason we do not see the aura is that the ability to do so has been taken out of us."

THE TEN SENSES

That was the answer. Kote explained that we have lost five of our senses. The senses of our imagination, five of our *ten* senses, have been removed from our education. The first five are the normal ones. Then you have the imaginal senses. The sense of self-healing—when you cut yourself your body heals. The sense of self-destruction—if you always create problems for yourself, you will break yourself down and cause wars, illnesses, whatever. The sense of penetration—to be able to penetrate other levels, other worlds, other dimensions. The sense of perception—to be able to see and understand what you are perceiving in those other worlds. And the sense of revelation—to be able to use what you have perceived because it has been revealed to you. These are the ten senses the Chumash work with.

As the boat motored along the Amazon, I had the realization that it was the five senses of the imaginal realm that shamans used in healing. What I still hadn't grasped was how this connected with the least-action pathway concept of healing.

A NEW PRINCIPLE OF LIFE

The sun had set more than an hour ago, and the dark and cool air felt very chilly. I was feeling the tension that Heisenberg wrote about.[7] We drifted along the Amazon, but my mind was miles away. I was moved by the teachings of Candace Lienhart, the Druids, and the vision of the Chumash medicine people. There appeared to be some principles of life based on the mind that we have lost. These are principles of healing that seem to be universal. Western thinking has somehow cut itself off from these laws. Western medicine and science have taken a view that everything is fundamentally dead. That life somehow arises by putting these dead objects together. If these dead objects are cooked up in a certain chemical compound, life will emerge. This is a view that starts with the Greeks.

I realized that this view was mistaken. As I saw it, then, the Western mind-set looks at the body as if it were some kind of machine. My new view assumes that all matter is fundamentally alive. That life exists everywhere, even in this table or chair. Life cannot be explained just by mechanical action.

Mechanical action happens when life becomes unconscious. It goes "dead" in its thinking. It then becomes mechanical in operation. And yet, our minds had to create mechanical action by creating least-action paths. These paths, as I mentioned earlier, became the unconscious mind—the body-mind that looks after our survival. The danger was that we, more often than not, tried to lead our lives unconsciously by going through the motions.

We all have tendencies to make life mechanical. We want to make ourselves follow ritual which is blind and enclosed and separates us from each other. That is the process by which life's spirit becomes entrapped and no longer is expressed in everything.

Suddenly, I felt anger. Life is cruel. We seemingly can't help ourselves. We all fall into the addictive behavior of mechanical action whether we want to or not.

Why was I here? What was I seeking so far from home? I wanted to learn how to break out of my own least-action paths. I wanted to confront my own unconscious mind. I wanted to find out by direct experience whether spirits really existed. I wanted to experience more fully than ever before the "subjective" mystical world that Heisenberg wrote about.

[7]See reference to Heisenberg in the Bibliography. He believed that we, today, must maintain the tension between objective rational science and subjective intuitive mysticism.

Candace told me about her "spooks." They were real to her. The Chumash told me how important the spirit world and the imaginal senses were to them. I was beginning to realize that my own mind controlled more than I was willing to admit.

It's not just the mechanics but also the envisioning spirit that is important in the healing process. Healing existed in that middle realm of tension of Werner Heisenberg's vision: the realm between the "objective world pursuing its regular course in space and time" and the subjective world of "a subject, mystically experiencing the unity of the world and no longer confronted by an object or by any objective world." Maybe I would discover the spirit world with Don Solon. The shamans believe that spirits exist. But there appears to be no such thing in Western medicine. If such spirits exist, then I wanted to directly experience them myself and not just within other people as I had in Tarapoto. Little did I know then that my wish was to come true.

THE VIBRATION OF THE AYAHUASCA MASTER

After an hour and a half had passed, we finally docked at a small pier and made our way across some narrow wooden planking covering the jungle floor toward a large enclosed hut. I was shown to my room, where I deposited my belongings. Don Solon had brought with him all of his mixtures, including the ayahuasca that we were to take, and he and Jorge went to the large gathering room to prepare for the ceremony.

At about seven thirty we all gathered in the room and took our seats on mats placed in a semicircle around Don Solon. A chair was placed for him to sit in. With me were Ted and Lisa, who had been there since Saturday, and, of course, Jorge. Arturo also sat in on the ceremony but didn't take any of the brew, and left after about an hour.

Without much ado, Don Solon began pouring the mixture into a small cup and first offered the drink to me. After me, Ted and Lisa, and then Jorge, and then the maestro, himself, drank.

I drank the mixture quickly and found it much more bitter than in my previous three experiences. The cup contained probably less than four ounces. I had prepared myself by not eating anything that day and very lightly the day before. I had hoped that the light diet would at least allow me to see a little of the future and perhaps more. I sat down and waited. After everyone had drunk, Don Solon stretched and began to chant the icaro very softly and slowly. I could hardly hear him, at first.

Then I noticed the vibration. The whole room, a good thirty by forty feet in area, resting on block cylinders of wood to keep the hut off the jungle ground, was shaking. I wondered whether we were experiencing an earthquake. I had noticed that there were several dogs kept by the staff who were attending to our needs. When the dogs laid down and began scratching at the ever-present fleas that "dogged" them, I noticed that the room shook slightly.

But nothing like a dog could cause this vibration. Somehow the song of Don Solon was connected with the vibration. I began to suspect that Don Solon was the perpetrator of this dance of the room. I watched him as carefully as I could. There was a kerosene lamp glowing, and I could see his whole body seated in the chair; nothing was moving. I looked over to Jorge. He was decked out in a hammock. No one else was moving.

I turned to Ted. "What is that vibration?"

Ted said, "He is causing it."

"How?"

Ted shrugged his shoulders. (Later, the next morning, I went into the room and took Don Solon's chair. I tried to shake my knees and cause the room to shake. I definitely felt a vibration, but not through the whole room. Mine was much weaker in comparison, and besides I had to move my body to make it. Later I began to suspect that somehow Don Solon had discovered a resonance condition and that even small nearly undetectable body movements, if at the right frequency, would set the room dancing. I still don't know how Don Solon caused that strong vibration. But however he did it, I knew that vibrational energy was important in the healing practices of shamans.)

I then thought about what it meant to the shaman to receive the ill vibrations from a patient. Of course, the shaman would get sick. That was the shaman's sacrifice.

SACRIFICIAL ACTION

I had always viewed extraordinary healing with skepticism. Certainly whenever I had a healing experience performed on me in the past, I could always attribute the healing to my own natural health. Even though I believed it was possible to heal in the manner that shamans used, I never thought it would work on me. The reason was my education as a physicist.

Deep down inside of me, as I mentioned in Chapter 1, in spite of how I expressed speculative ideas about the new physics, I had a classical mechanical Newtonian machine running. This machine was my own mind. And because the majority of my experiences, as I am sure is true for you, could usually be explained via the normal operating mechanisms that appear to govern our lives, such as Newton's laws of motion, cause-and-effect reality, and determinism, I unconsciously must have viewed all parapsychological and paramedical phenomena with unconscious scorn.

I did remember the spontaneous healing that I had when I was suffering with sinus pain. Several years ago, my wife, Judith, and I had driven from our home in La Jolla, California, to Ojai to visit with Joan Halifax, who runs a spiritual and healing retreat there. I had felt well enough when I left, but by the time we had arrived, my head was splitting, and I told Joan and Judith that I had to lie down. Joan introduced me to a young Native American, who said if I came with him at that moment he would heal me straightaway. Joan and Judith encouraged me. I, of course, was skeptical and just wanted some pain-reliever pills.

The Indian took me to a quiet place and asked me to do some exercises which involved stretching of my body. He used long poles that are usually used to make tepees. He asked me to hold one end of a pole and put the other end into the ground. He then had me stretch my body, leaning into the pole, holding that end as a lever and a support. I felt excruciating pain. He told me to yell and he pushed on my body so that it would stretch even more. He chanted a song as he stretched my aching muscles. I felt as if every muscle in my back and neck and arms were about to burst. I screamed.

When I recovered, he asked me how I felt. I said my muscles were quivering and I was tired. "No," he said. "How is your head?" Suddenly I remembered that I had forgotten my head. When I took awareness of it, the pain had vanished.

My body was a sacrifice to my head pain. By enduring physical pain in my muscles, I was able to release my headache. Perhaps the head pain was released to the universe that way. And perhaps that was the secret to how shamans got rid of their patient's illness; they, too, made a sacrifice to the universe.

As the first hour continued, Don Solon's song became stronger. I marveled at the number of ayahuasca journeys he had taken throughout his lifetime. Each one was an ordeal, I am sure. I definitely felt that I was in the presence of an expert. In his preparations I noticed that he blew strong tobacco smoke into his mixtures. I saw that tobacco was to play a more important part in Don Solon's ceremony than in any others I had experienced.

About forty-five minutes after taking the brew, I began to have a familiar feeling at the onset of visions. I saw many geometric patterns that seemed to be in phase with the icaro. None of these patterns were pictures, however. Soon I began feeling *mareado*. Then the picture show began. I was so dizzy that I immediately lay down on my back hoping that I would thereby hold the ayahuasca down and not throw up.

THE SPIRITS OF THE AYAHUASCA

I first saw a vision that completely mystified me. I saw two tiny deep-brown-skinned people wearing white wraparound knee-length skirts. They were quite small, appearing to me to be less than five feet tall. They were also very thin, but they didn't appear emaciated. Far from it, they seemed quite strong. Their features were aquiline, fine-lined. They did not appear to be Indians, even though their coloration was clearly not white.

They walked with regal or princely bearings. I felt that I was watching members of some ancient royal court. Except for these wraparounds, these two men were naked. They were totally hairless except for their heads, which were completely shaven except for long tufts of hair somewhat like Mohawk haircuts, which stretched from the nape of the neck to the top forehead region. That hair was nearly a foot long, shiny black and spiky, pointing straight up from their heads like radial spokes from a wheel's hub.

They were walking by a black iron-grated fence, which I could see through. Through the grates I saw an immense pool and very modern architecture. Yet when I looked again, the architecture, although looking modern, really wasn't. The shapes of the buildings appeared geometrical, with a wide range of slanting lines. They didn't appear pyramidal, but I had the impression that the lines of the buildings were not just vertical and horizontal. More triangular in shape. The buildings were white and gleaming in the bright sun. The sky was a deep rich cloudless blue. They seemed to hardly notice me, but they did see me. They walked through the gate attached to the black iron-grated fence. It reminded me of a typical gate in front of a pool at a modern hotel. They beckoned me to follow them into the pool. Then the vision vanished.

I WAS MISTAKEN

Immediately a series of nondescript patterns and faces passed before me. I saw horrible grotesque masks. However, they didn't frighten me, and as I was lying down, I didn't feel dizzy or faint. In fact, I felt well and had believed that I would be able to hold the brew and not throw up. I thought I had it beaten. I was mistaken.

Soon the next part of Don Solon's ceremony began. This was the healing icaro. Jorge told me to get up and kneel at Don Solon's feet. I got up very reluctantly and immediately felt woozy. I knelt in front of the seated Don. He asked me, "Are you mareado?" I said, "*Si.*" Then he began to work his magic on me. The room had not stopped vibrating during that whole time. Kneeling before Don Solon, I felt the vibration even more than before. I had to lower myself. I sat down with my legs outstretched. Don Solon lit a cigarette. He began to blow the awful-smelling tobacco, from deep within his lungs, on my head, down the back of my neck, down the front of my chest, and all over my body. It nauseated me completely.

I was exhausted, as if all of the steam had been taken out of me. I was overwhelmed and began to feel sick. I crawled over to the open side of the room, put my head over the side, as if I were on the deck of a ship, and threw up on the jungle floor five feet below me. I seemed to give up enormous amounts of fluid. I had only taken four ounces. Where did all of that come from?

I returned to my position at the Don's feet. He continued administering the tobacco-wind and song. I then felt well enough to look up at the cause of all my misery. What I saw amazed me. Don Solon became an old tree. His skin kept changing from human to bark. He was a forest spirit himself. The tree was dry and old. His voice seemed to be coming from the earth, and it appeared that it was the voice of a long dead being singing. He did not seem to be a light spirit at all. Very dark. Yet I did not feel afraid; his was a benevolent spirit.

Don Solon's singing had not ceased even after I returned to my place, and he began his healing with Ted. In fact, the ceremony lasted for six hours, and Don Solon's songs never left the air.

A GROWING RAGE INSIDE OF ME

The ceremony ending was marked by Don Solon's speaking voice. For some reason I was feeling very angry. Don Solon, ignoring my anger, arched his back and asked me to tell him what I saw. He wanted to know what I had experienced from the time I took the brew. I told him, "My first experience was the bitterness of the brew, and I was surprised at how small the portion was and yet how potent it was." I then told him about my visions of the little people. I said, "I am still seeing them now. I am in awe of the power I felt coming from you. The experience nearly overwhelmed me."

Don Solon smiled and told me that he knew these spirits. He had seen them many times before. I asked him who they were. He told me that they were the spirits of the vision vine. He said that this was a good vision and that they were welcoming me to their world. Ted and Lisa also said they had a similar vision, although it wasn't quite the same as mine. They saw a lake instead of a pool, and they didn't describe the spirits.

But my anger had not subsided. I then confessed my feelings. I noticed that I was quite calm telling everyone what I had felt. I told them that I had felt as if no one wanted me there and that I was somehow in the way of Ted and Lisa's experience.

I hadn't realized at the time that what was coming through me were past experiences of feeling unloved and unwanted as a child. I had transferred my feelings about my father to Ted. I never felt that my father ever loved me. Many years before, a well-known psychic and dear friend, Carol Dryer from Los Angeles, had told me that she had seen that my mother and father had made a pact at some psychic level that they would have each other and the great love they had for each other, if they were willing to put up with my birth. I was going to be a difficult and bright child. Certainly Carol's insight rang true to me.

After my confession, I immediately felt great relief. Ted was looking at me with shining eyes. He came up to me and embraced me and told me that all he felt for me was great love. I relaxed in his arms like a child and began to weep. I didn't know then what had brought this all up. Jorge then said, "It is okay. The ayahuasca opens the unconscious, and whatever is there, controlling our lives, surfaces. I have seen such anger before."

I felt tired and got up, saying good-night to all, and went to my bed. After taking off my clothes, I covered the bed with the mosquito netting, crawled in, placed a sheet over my naked body, and after around fifteen minutes of tossing and turning, I fell asleep.

LESSONS FROM THE JUNGLE MOTHER

When I awoke the next morning, I began to look at my surroundings. It was Thursday and I found myself in one of the many open rooms of the large jungle house consisting of huts connected by wooden plank walkways.

It was primitive to say the least. There was no electricity or running water. The toilet consisted of the typical outhouse 100 feet to the rear of the main structure. It consisted of a raised plank board with a one-foot diameter hole covering a five-foot-deep dug pit.

I bathed and washed that morning in cloudy Amazon River water. I simply took a bucket to my room and poured it over my body. It felt wonderful. Even though we were practically spot on the equator, it was cooler here than in Tarapoto.

After dressing, I made my way to the dining area. Don Solon, Jorge, Ted, and Lisa were already up and about and were eating their breakfasts.

I asked Don Solon, "Were there any spirits present in the room last night?"

"Yes, of course."

"What were they?"

"They were spirits of my ancestors and spirits of the plants."

"What did they look like?"

"Every plant spirit has its own image and its own quality."

"Did they notice my presence?"

Don Solon smiled at me. "They were very curious about you. They were all around you and they were admiring you. When I sang my song for you, the spirits entered your body and cleansed it."

"I noticed that I became even more dizzy after you blew tobacco smoke on me."

"The tobacco, too, has its own spirit. The tobacco spirit is a little blond man."

"Why did it make me feel so ill?"

"The vomit comes when a purge is necessary. There was something in your body, your stomach, that no longer was okay with you. The stomach acts as a way to cleanse your body and its spirit."

"But was just my fear causing my illness?"

"No. You needed to clean yourself of past accumulations. This is how the body does it. It cleans itself from both the inside and the outside this way. Both the mind and the body. Anything that is not in accord with your purpose is removed."

As we talked I noticed the sounds of thousands of birds. I could also hear the boats going back and forth along the Amazon River.

Soon Jorge and Don Solon left. Don Solon wanted to return to his family, and Jorge had a daughter living in Iquitos. They both were due to return to the hut Friday evening for another ceremony. I walked with them to the pier and waved good-bye.

That evening, I retired early and began to think about what had happened to me. Certainly over the past few years I had been struggling with some health problems. I was beset by chronic sinus headaches and a chronic acidic stomach. No wonder I had felt so ill.

I then recalled a mild schizophrenic episode I had felt during the ceremony. I had heard my self split into two beings, two voices. One voice, my small self or my ego, was attempting to control the experience, make sense of it all. I remembered that it was the same voice that I had heard during my very first ayahuasca experience—a nagging one that complained and suffered. While it spoke, I was very upset. It kept saying, "See, what have you got us into this time. Maybe you're going to die. You're as sick as a dog. It's stupid. How can you be so stupid?"

But then I realized that this was just one of my many voices, a little like Jiminy Cricket talking to Pinocchio. So I said to my nagging voice, "I know it already. But stop complaining because it isn't doing me any good."

Then this voice, feeling reassured, became curious. It was asking questions. "Ask the spirits of the ayahuasca to show you what an atom looks like. Ask them to take you to the future. Don't just lie there, do something!"

But there was another voice, more experienced, deeper, slower, and wiser. It answered the first voice's inquiries. "Take it easy. Relax. Observe the journey. Don't even try to control it." I was in the middle of these voices, so I guess I had become three people. A teacher, a student, and a witness to the whole show. The teacher was encouraging my student to keep quiet and just pay attention, witness whatever happens. No judgments or thoughts were needed.

I had also undergone a crossover of my senses: sounds turned into visions. (You'll recall that I discussed crossovers in Chapter 5.) These crossovers, induced by the vision vine, were alterations in the neural pathways in my brain. They were harbingers telling me that a new way of seeing and sensing was possible. I was beginning to realize that they were actually inducing new least-action pathways in my neural network—something I now suspected was vitally necessary for healing to occur. Then I realized that I was ill and that I had come here not only to find out how shamans heal but to be healed myself.

At one point in the ceremony when I was feeling quite sick, I paid very close attention to Don Solon's song. The melodies were quite simple and very haunting. My whole body seemed to resonate with the song. I closed my eyes and saw the colors of the song. Each change of tone, the rhythm, and the loudness with which I participated caused a visual show.

LYING IN THE JUNGLE MOTHER'S WOMB

As I lay in bed, I wondered if I was still under the influence of ayahuasca. There were weak remnants reverberating through my nervous system. I could still see colors, and at times visions of faces would pass before my eyes. I realized that I was falling asleep and that I was witnessing a familiar hypnagogic state—visions that come on in that twilight of consciousness between wakefulness and sleep. I was feeling strangely comforted by the sounds of the birds and the nearly perfect temperature. As long as I lay still, I could barely distinguish the warmth of my body from the humid and warm air of the jungle.

You can't hate the jungle. I don't think you can like it either. The jungle is really the earth mother's womb, and it must be respected. All life flourishes here. As in the womb, life has its way. Evolution takes its path and life *is*. Life leads to life. And like the path of a developing embryo, as it grows and thrives other life-forms live and then die.

The jungle is accepting. Like a good mother, she feeds her life-forms and she eats her life-forms' excrement. Everything is in balance. The jungle gives and it accepts. It is delicate, and at the same time it is very strong. The jungle is like Don Solon. "He is old, but he is very strong."

I got up from my bed. It was late and everyone had gone to bed. I took my flashlight with me and walked the planks from the sleeping quarters to the large ceremony room. I listened as I walked and I heard the sounds of fluttering wings, some passing quite close to my head. I turned on the flashlight and looked up. There were perhaps fifty bats hanging from the rafters. Strangely, I hadn't noticed them the night before. They were there. They are there every night.

As I stood there in the cacophony of the night birds watching the bats, I recalled a brief moment I had with my guide, Franco. Suddenly I remembered that I had spoken with him without using any English for at least three minutes. It seemed so innocuous to me that I barely even thought about it. The ayahuasca must have loosened my fear of speaking a foreign language.

I considered the jungle-mother and the ayahuasca-milk she fed me to be my teacher. It was evident to me that I had a person inside of me that wanted to know and speak Spanish. When I let him come forward, I found him to be very young and sweet. He was no more than a teenager, perhaps even a preteen. That was the person who hung onto Spanish. My English-speaking person was much older, much more sophisticated. He usually ran the show. But in Peru, he didn't understand anything. He had to take a back seat, and in time he had to ask the teenager, "Just what the hell is going on here?"

The jungle had grown very quiet. I returned to my room, and crept under the netting. I lay there listening. I relaxed and began to realize that I had learned some of the secrets of shamanic healing.

Of course, at that time my thoughts were somewhat confused. It is only now, after the movie, while I am waiting here in my Lima hostel room, for the flight back to the United States, that it all becomes clear to me.

The first revelation was that the patient was being healed by learning to extend his consciousness into the five senses of imagination as the Chumash medicine people taught me. This was possible through the creation of new least-action pathways in the neural network of the brain and nervous system. I recalled those senses again:

. . .

1. *The sense of self-healing*—when you cut yourself your body heals. Our bodies have this sense and we can learn to recognize it. In other words, it is possible to recognize when you are healing yourself. I remembered that all of my healing took place with a great sense of emotional release. But to fully realize this sense, I had to literally enter another way of "seeing," again a realization of a new least-action path in my brain. I had to reach into the mythic element of my own being. I had to release my controlling mechanical mind and, in a sense, "go crazy."

2. *The sense of self-destruction*—if you always create problems for yourself, you will break yourself down and cause wars, illnesses, whatever. I was certainly aware of this sense. It overcame me in times of depression, times when I felt my self-worth was extremely low. During this time I would feel very much alone and isolated, regardless of where I was and whom I was with. It was only now that I recognized that this too was only another way of "seeing"—another way in which I "went crazy."

3. *The sense of penetration*—to be able to penetrate other levels, other worlds, other dimensions. I had only a dim awareness of this sense, but I recognized that I was able to penetrate into the other way of "seeing" when I noticed, for example, extraordinary synchronicities in my life. These were messages from the mythic time of my own life. I would feel this only dimly at such times as when I fell in love or when I became aware of deep intuitive ideas. My ability to penetrate depended only, again, on my willingness to release any preconceived ideas about what I thought the "real" world consisted of.

4. *The sense of perception*—to be able to perceive events and the surrounding world in a different light. Once I had entered into this other way of "seeing," I was able to perceive normal events as extraordinary. Here, again, once I was able to release my mechanical mind, *any* event could be perceived in an extraordinary manner. To fully realize this sense, one had to trust one's intuition even if the concepts one realized were totally "off the wall." There were many times I could remember when I knew I was ignoring this "inner" sense. And invariably I made mistakes when I did.

5. *The sense of revelation*—to be able to understand what you have perceived in these other worlds. During those times when I had perceived events as extraordinary, I was able to grasp the information I had received and use it when my mind once again entered into mechanical thinking. In other words, I was able to bring my feeling-intuitive self into my thinking-sensation world. I was able to put into words my understanding of what was revealed to me.

. . .

My second revelation was that shamans working through these other senses can work miracles, such as the manifestation and transformation of matter, particularly in the human body through the creation of new least-action pathways in the physical body of their patients. How were they able to do this?

They did this by finding meaning in personal events in the life of the patient that would often have been dismissed as meaningless. Once two seemingly insignificant events were found to be meaningful, a least-action pathway would ensue between them. Often the events would be found to be significant when seen from an intuitive-feeling point of view rather than a thinking-sensing point of view or from a mythic viewpoint rather than a literal one.

The reason shamans can seemingly work miracles is simply because they believe they can, and they base their beliefs on their knowledge of archetypes from the imaginal or mythic realm. Western scientists do not believe in miracles. In a certain sense, neither do shamans. What they do is perfectly natural to *them*. They don't call the appearance of plant spirits or spirits from the dead miraculous. They bring them forward using songs and other techniques such as imbibing ayahuasca or striking a drum repeatedly. When these spirits appear, the shamans are not surprised. They can repeat the effect anytime they wish.

In a similar manner, the physicist brings electrons forward by using refined technological instrumentation. And, of course, no physicist is surprised when electrons appear. You must remember that electrons were not even discovered until 1896. The first discovery showed that electrons were tiny electrically charged particles. But later, electrons were discovered to be waves. Thus electrons had two contradictory behaviors. Both had to be true. Thus when a physicist performs a certain kind of experiment, an electron wave appears. If he performs another kind of experiment, an electron particle appears. To a skeptic who didn't believe in electrons, this would be evidence enough to dismiss the whole idea of electrons as patent nonsense.

The third revelation was that shamans do "heal" in the full sense of the meaning of that word according to our classical Western medical practice, and their healing abilities could be explained by the emerging new paradigm of quantum physics. They healed by initiating patients, causing them to vibrate in resonance with themselves. By transferring vibrational energy from the patient to the shaman, the patient would heal.

A key insight is vibration and vibrational patterns. The shaman is able to produce a healing vibration in the patient's body. When the patient "tunes" to this vibration, he is healed. That tuning probably has something to do with a vibrational frequency of a quantum wave of probability. The frequencies of these waves are related to the energies of the particles, which tend to manifest where the wave interference patterns are thickest. The lowest-frequency waves would have very long wavelengths.

These waves are probably waves of nonvisible light, possibly in the long radiowave part of the spectrum. I believe that the electromagnetic spectrum is involved in shamanic healing. I believe that the shaman is able to alter these waves in ways that we have lost. You'll recall that in Chapter 5 I mentioned there was good evidence that shamans probably were more sensitive to the nonvisible radiation discovered at sacred sites.

In brief, the world of the shaman and the world of the physicist overlap in the imaginal realm. The electron, the quantum wave of probability, the photon, are the same as the archetypes of the shaman. The electron has complementary attributes. Probably the spirit is another form of complementarity.

Thus there is a deeper complementarity in the human soul. The human spirit operates in the feeling-intuitive world. The material human being operates in the thinking-sensation world. The thinking-sensation world is ruled by chronos—our ordinary clock time. The intuitive-feeling world is ruled by mythos—the sense of timelessness. It is only by realizing that both worlds are necessary that a person becomes healed, all one again.

And there was something else required. In order to heal, I had to be able to enter into other worlds as easily as I walked in this one. All of the lessons I had learned pointed to the reality of these other worlds. They were accessed through the imagination. I had to learn to "see" in the imaginal realms as well as I apparently saw in the physical world.

Then I realized that the imaginal realm could be reached through our dreams and that I had already gone there several times.

8

Other World Experiences

I was in Lima watching the movie and I was still thinking. It was the mythic world that contained the power of healing. But what did the mythic world consist of? The story of Miguel paralleling my own was revealing this to me. At some mythic level, I was on a hero's journey—a search for truth and meaning in my life. At the physical level, I was here in Lima. I was really living in two worlds.

First there was the world of objective reality—that world that we believe is composed of objective facts. It is an apparent world of hard reality that no one seems to dispute. Science is the study of that world. Science hopes to show that all phenomena can be explained in terms of objective repeatable experiences. It sees the universe constructed of objects. The game of science is to map all experience in a temporal order: fact following fact, cause leading to effect. The objective world that science seeks to find is the world of time: chronos.

But there was another world, a world of the mind. Our minds attempt to map experiences, making them appear as objective facts—grist for world one. But in the second world, we also have our imagination.

This second world was mythos. It consisted of what we believed was possible, out there. We know that it is impossible to completely map the first world without input from the second world. For example, we have never seen a subatomic electron. We believe that it is real. By choosing to see the electron in a physics experiment, we are actually doing no such thing. We are taking ideas from world two and applying them to world one. But the inevitable complementarity of the wave-particle duality of the electron forces us to give up the task. We cannot know the electron as an objective bit of out-there matter. Its properties depend on how we look at it and what we choose to look for. We alter its properties by our probing into its nature.

It was once said by Sir James Jeans, the British philosopher, "The Universe begins to look more like a great thought than a machine." And the more I journeyed into the shamanic world, the more this seemed true to me. It was clear to me that shamans use powers to deal with a universe we in the Western world tend to dismiss as imaginary. But that tendency was gradually vanishing, since the discovery of the hidden world of the atom and subatomic matter made physicists realize that it was impossible to entirely ascribe objective properties to ultimate matter and energy.

Later that evening I recalled my sixth hypothesis: Shamans enter into parallel worlds. It was now apparent that these other worlds were found in the imaginal realm, not just in the physical realm. Originally I believed that these other worlds were parallel physical realms, just like those predicted by quantum physics. Now I wasn't sure that this was right. Instead I began to see that the world of mythos was parallel to the physical world in a different sense. Somehow it was a world that eluded the uncertainty principle. It was a world that was determined in a way that my quantum physics training could not fathom. Somehow when shamans dealt with that world, a degree of control—and, yes, even destiny—that wasn't possible in the physical world was manifested. That shamans could tune into that destiny and that prospect, for some reason, frightened me.

It was becoming apparent to me that the real power of shamanic physics was the ability to alter matter by altering the world of the mind. This meant that shamans must have the ability to enter the mind world *and* they must be able to find their way back. I then began to remember how I discovered the mythic world. I was back in the jungle waiting for Don Solon and Jorge to return.

I awoke at around seven, having slept well. My headache was gone and I was feeling very well. I made my way to the river, jumped in, swam and bathed. It was Friday morning. I had been in Peru, now, for ten days, seven of them in the *sylva* (jungle). I felt refreshed, got out of the river, and made my way across the planks to the hut.

It was a Friday, and remembering the Jewish traditions, that meant that this night would be different from all other nights. I was to take ayahuasca again with Don Solon.

I breakfasted on fruits and tea. I was not planning to eat another thing that day. Franco asked me if I wanted to take a canoe ride to another village and meet some Indians. I said I did. We then walked for a while along the jungle-river edge, and I felt transfixed. I was feeling as if I were walking in a dream. I once more turned my thoughts to my past. I remembered a dream I had that, in a sense, really started me on this journey. At the time, I hadn't thought so. It took place in England more than fifteen years earlier.

MY DREAM JOURNEY TO A SHAMANIC WORLD

I was then living back in London and had returned to my office at Birkbeck College, next door to Professor David Bohm. Bohm had often come into my office to discuss his ideas about physics and consciousness with me.

Professor Bohm had a rather hypnotic way of presenting his ideas, even though I didn't fully understand them. I would often return to my flat in Shepherd's Bush filled with Bohm's thoughts and wondering about what they meant.

One evening, after a particularly engrossing afternoon discussion with him, I found myself growing quite tired very early. And yet, I was quite agitated with his ideas. I fell asleep quite quickly after going to my bed, joining my lover and housemate, Nancy, who had already gone to bed herself. Nancy was very mystical, and between us there were many magical moments of experiencing a oneness in consciousness. Perhaps it was her presence that caused the following remarkable dream experience to occur.[1]

About a quarter to four in the morning, I had awakened. My mind was full of quantum physical equations describing parallel universes and alternate realities. I rose from the bed and went into the dining area of the flat to try to write down what had been filling my mind. After about a half hour of writing, I suddenly became completely relaxed and very sleepy. I went back to the bedroom, trying not to disturb Nancy as I crept back into bed. As soon as I put my head down on the pillow, I fell into a deep sleep.

However, I never quite lost consciousness. In fact, the room around me began to spin, and I felt myself descending as if I were passing from one layer of reality to another. After a while, I thought I had awakened back in my bed again, but was surprised to find that I was still asleep. The room I had awakened in was a dream room. I then realized that even though I was asleep, I had actually awakened in my dream. I didn't know it at that time, but I was undergoing a typical shamanic journey of my own.

[1]I described this sequence in my earlier book, *Star*Wave: Mind, Consciousness, and Quantum Physics*. I am including it here in more detail.

According to Michael Harner and Mircea Eliade,[2] known experts on shamanism, "a shaman is a man or woman who 'journeys' in an altered state of consciousness usually induced by rhythmic drumming or other types of percussion sound, or in some cases by the use of psychoactive drugs." Carlos Castañeda, in all of his books,[3] points out that a sorcerer [like a shaman] journeys to "nonordinary reality" in an altered state of consciousness. Harner calls that state of consciousness the *shamanic state of consciousness* or *ssc*.

Although I had not undergone any drug experience, nor had I undergone any drumming sounds at all, my work with Bohm's ideas the weeks preceding, I am sure, qualified by drumming into my head to alter my normal waking consciousness.

My sleeping journey has been labeled by today's dream researchers as a form of *lucid dreaming*. At that time I didn't really know what was happening to me. My experience was one of falling and spinning and then awakening. Whenever I awakened, I saw a full-color world of things happening, only they were passing before my eyes very quickly in a dizzying manner.

Now when I say I fell asleep, I mean more than you might think. I felt myself falling down a deep and dark well or tunnel. Yet, every so often, I would stop falling and find myself involved in a scene, as if I were an actor suddenly appearing on a stage. These scenes just appeared and I was enmeshed in them. I was not just an observer but was actually "there." Quickly the scene would change and I would find myself in yet another scene, entirely different from the one I had just left. These scene changes would happen so rapidly that I felt that I was descending from one universe to another, slipping through time and space just as a small pebble slips through a distorted woven mesh of a fabric.

As I descended, I became more and more aware that I was dreaming. It was dawning on me that I was both snuggled cozily in bed next to Nancy and slipping through space-time in a dream of uncanny proportions. It was as if my awareness was split in two. To my great surprise I was conscious and I was asleep. What a contradiction! How can you be asleep and consciously aware at the same time?

As I mentioned, when I found myself awakening, I was shocked to discover that I had not actually awakened at all: I was dreaming that I was

[2]See "What Is a Shaman" by Michael Harner, in Doore, Gary, editor. *Shaman's Path: Healing, Personal Growth, and Empowerment*. Boston & London: Shambhala, 1988, p. 7.

[3]See, for example, Castañeda, Carlos. *The Power of Silence: Further Lessons of Don Juan*. New York: Simon & Schuster, 1987.

awakening and I knew it! No sooner would I realize that I was still dreaming than I would awaken once more from the dream to dream that I was awakening once again.

I soon realized that I had some control of my dream. I could awaken for real and get up from my bed, or I could descend to any other new universe layer I wished and experience my dream consciously. At that point I decided to explore where I had landed and instantly found myself in the strangest place I had ever been in. I began walking silently and saw that I was in a quiet, beautiful countryside of bright green, grassy, rolling hills. Soon I "heard" voices and saw a large group of people just ahead of me.

My entrance into the group stirred no response. I began to look around at the faces of my new associates. I must point out how unusual this was for me, because at any moment I could "remember" myself sleeping in bed at home in Shepherd's Bush, London, England. "I" was where "I" was and "I" was home at the same time. This experience of remembering was exactly the same as when you think back to a past experience, the only difference being that in normal waking consciousness, you can't "return" to your memory. In my altered, or lucid, dream state I not only remembered my sleeping self, I knew I could return any time I wished to.

Soon it became apparent that I would want to do this quickly. I had come upon an astral level where I felt the people there had previously committed suicide. Aside from this, another reason I was thinking about going home was the bizarre physiognomy I was suddenly gifted with. I merely had to look at a face, any face, and I "saw." More than seeing, I knew. The face would undergo a series of transformations, each change revealing a new fact. I couldn't look too closely because, frankly, I was frightened by what I saw. The faces were normal when looked at quickly with glancing images, but when examined for any length of time they became grotesque masks with great striations of contorted pain lines, hideous peelings of unfolding skin layers, and throbbing nerve threads all pulsating on raw and reddened skin.

Just then the danger became apparent to me. I noticed a woman sitting on a wall, smiling at me. I was surprised at this, because no one had been paying me any attention at all. So why did she? I then "heard" her.

"Where are you from?" she asked. "Who are you? Why are you here?" she continued in an overly friendly manner.

I "said," "I am not really dead. I am not really here, but back in bed on earth."

"Oh you are, are you?" she said as she now approached me.

"Yes, I know where I am, and I can return home anytime I want to," I replied.

"You can, can you?" she replied with increasing interest that made me grow more suspicious. She was coming closer to me, and I was becoming more frightened with each step. This was, after all, my first trip here, and I didn't know what danger I might have been in.

Then I looked into her eyes. I don't know how to describe what I saw, but her eyes began to spin. They appeared to me as rotating pinwheels of spiraling colors. She was now too close for comfort, and I was growing dizzy looking at her. I knew that I had to leave, and I exercised the "leaving ritual," the only one that I knew would get me out of there fast. I yelled as loud as I could.

I awoke in bed next to Nancy, thank God, this time for real. It was now nearly 5:00 A.M., and I am sure that Nancy wasn't too happy to have me just pop up in bed talking a blue streak. Not only was I wide awake, I was fully conscious and quite lucid and gregarious. Loudly, I said to her, "Nancy, wake up. I must describe this dream to you before I forget it."

Nancy, hardly believing her ears or her eyes, was rudely being shaken from a deep sleep of her own. And dazed, but sportingly understanding, she listened to the story of my voyage.

It was very important to realize that this "dream" was not just an ordinary dream. I was fully conscious not only during it but in the transition from the astral level to my bed. My yelling was soundless in that astral realm but became real sound in the physical plane of my bedroom. There was no need for coffee. There was no sleepiness, nor did the dream fade from memory as I became more awake (as most ordinary dreams do). It was simply a matter of recalling actual events in the same manner as you would recall events of the morning after having your lunch.

I hadn't really been asleep. I hadn't been lying in bed daydreaming or fantasizing. It was too real for an ordinary dream.

Although I have had several waking or lucid dreams since then, I have never returned to the astral plane of suicides again.

Shortly after the dream, I realized why that astral level of reality existed. This astral level was a holding station of persons who had recently taken their own lives. These people had committed suicide on earth and were waiting to reincarnate—to return to earth and be reborn.

In order for them to return, they had to be acceptable to all the "normal" nonsuicidal souls they would come to share a body with. That is why they were here: to await humanity's decision.

All of this, of course, didn't dawn on me at the moment I had the dream. Years later I realized that each of us is a universe of souls, not just a single soul journeying from here to God knows where. As was taught by the

Buddha, we are all questions of compromise—villages of souls and desires. Each of us is a universe of past lives, and some of us living now owe a debt of gratitude to the others for allowing us to live again. These suicides were the astral-level component, the parallel universe level of reality, of past failures in life. We all have in us the lives of past failures, criminals, rapists, murderers, saints, and sinners.

However, here in Lima, watching the shamanic journey movie and Nexy's confrontation with the spirits that are haunting her, I still wonder about my dream. Why should I have been taken to a plane of death, and suicidal death at that? Perhaps it was a warning for me. Perhaps it was a vision of a future suicidal tendency I might develop. Or perhaps it was a message to me that I was becoming aware of my own shamanic power.

OUT-OF-BODY EXPERIENCE

How does one become aware of one's own shamanic power? I was seeing a way, although it wasn't as I had expected it. The key is to recognize in one's own experiences the presence of shamanic incidents. These events are always powerful, but when they occur we often tend to dismiss them as foolish, frightening, or coincidental without any further merit. These experiences can be thought of as bleed-throughs from a world that is beyond our immediate sense of the common world we all live in.

In the past years of my life, I had my share of bleed-throughs. These were definitely points of contact with other worlds. As I sat watching the movie in Lima, I remembered. And as I remembered, the vision grew stronger.

My mind went back through time again. It was back in February of 1975, when I left England and returned to the United States.

My first stopover took me to Ossining, New York, where I had the opportunity to visit with Dr. Andrija Puharich. I had met Andrija ten months earlier at the May lectures in London. He had told me that he had a group of "psychics" all living with him at his estate near Sing Sing Prison in New York State. Dr. Puharich had been researching magic in the world for more than a quarter of a century. He had made several remarkable medical inventions, but still had the time and energy to devote to his research into magic. He had spent some time in Hawaii and had been "ordained" as a kahuna—a shaman following the Hawaiian tradition. He described some of his study of shamans in his book, *Beyond Telepathy*.[4]

[4]Puharich, Andrija. *Beyond Telepathy*. New York: Doubleday Anchor, 1973.

Puharich had returned from Israel a few years before, where he literally had discovered Uri Geller, a young Israeli with remarkable psychic powers according to many who have witnessed them. Geller, according to Puharich, was capable of traveling from one place to another without the passage of time. He also had the ability to discern information incapable of being detected by other means. However, in spite of Geller's powers, there was a huge amount of controversy about him.

I had been invited to spend as long as I liked with Andrija, researching his work with psychic phenomena and attempting to draw a connection between his discoveries and my work in quantum physics. I was curious to discover if Andrija had any secrets that I could discover. Was all of this New Age magic just so much hokum? Or was there some hidden truth to it all? With an open mind, I decided to spend some weeks at Puharich's home in Ossining.

Puharich's spacious home was on a large expanse of ground surrounded by northeastern greenery. There wasn't a day that passed without incident, since the house was full of unusual people passing in and out nearly every hour of the day and night. Puharich had, in his basement, a copper-screened room where he had been conducting experiments to see if telepathic communication would pass through metallic copper—something that ordinary electromagnetic waves would not. A number of "psychics" were being investigated, and I was exposed to them on a daily basis.

I can't say that I discovered anything that would lend credence to their abilities. (There was a young woman who claimed that I could read her mind—but she didn't have much success reading mine!) However, one evening, a few weeks after my arrival, I experienced my second journey in altered consciousness. Just before I had retired, I had a vigorous discussion with Andrija and some of the people in his group. Afterward, I felt very tired and went to my room to sleep.

I awoke to find myself floating out of my body. This time I was prepared to investigate what was happening to me. I noticed that I had risen approximately five feet above my sleeping body below. I then encountered another life-form and wondered what it was. Later, after I had awakened, I looked above my bed and noticed that there was a large plant hanging from the ceiling. In my altered-conscious state, the plant had appeared quite different to me. It seemed to be alive in a more animistic state than just a hanging plant.

When I fell back to sleep, I began to journey again and was surprised to find myself back in England driving a car down the Bayswater road. I knew that I was asleep back in New York State and at the same time, in a different form, I was driving a car in England. I decided to conduct an experiment. I deliberately turned the car so that it would collide with oncoming traffic! I don't know why I chose to do such a violent thing, but the event was amazing to me. My car crashed into another car, and my car began to turn over as a result of the collision. I then felt my body turning and I began to awaken in my own bed. As I came out of the dream, I noticed that I didn't quite fit into my body. I felt that my dream hand was separated from my real hand, and I remember the feeling of trying to slip my hand into my hand, just as it would feel to slip your hand into a glove. Gradually I made my way back into my body, and after a few moments, I awoke fully integrated.

At that time, these out-of-the-body experiences were impossible for me to explain. I tried. I kept thinking that they were no more than some form of hallucination—whatever that meant. But the word "hallucination" does not explain the experience. It simply dismisses it by covering it in scientific jargon.

What is a hallucination? Some elements are objectively understood. First of all, you have sensory information that tells you that what you are experiencing is real. You can see the "thing" happening. You can "feel" it by touching it. You can "hear" it, but in my case, the hearing always occurred in my own mind. I could remember at times, especially just before falling asleep, being jolted awake by the sound of a person's voice; sometimes it was a person who had died years before. This has been called an "aural" hallucination. But these "aural" hallucinations were never accompanied by other sensory information.

Sometimes you can even taste or smell the "thing." I have now come to realize that a real shamanic experience is always a combination of at least two or more of the common senses. Most often the sense of sight is one of the senses invoked, but not always.

My work in parallel realities had convinced me that hallucinations are altered states of awareness. But awareness of what? In the discipline of physics, there was supposed to be only one world, the real world made of matter subject to whims of energies that pushed and pulled on particles. The world was a dead one animated by energy. But quantum physics, although initially subscribing to that picture, was pointing to a different view. The world can never be completely known in terms of matter and energy. Matter is not fully material; it is something else, and by altering the way we observe the world, we begin to see that something else. We become aware of other worlds.

In an altered state of awareness, such as my lucid dream or my out-of-body experience, I was actually witnessing other worlds. These worlds were hidden in our normal waking consciousness, because we have become conditioned to seeing in a certain way. This way developed as a respect for the inertial properties of matter. In other words, this way was necessary for survival. However, in our attempts to rationalize our experience, we have lost another way of seeing that is also necessary for our survival.

Shamans had that way of seeing at their disposal. They could travel to other realms, and now I had realized that I had been given that shamanic gift without really knowing it at the time.

When you are very ill, especially with a high fever, you can also drift into other worlds. Most of us have done this as children. I remembered my meeting with Chris Hall and Richard Dufton in Brighton and what they told me, last year.

DRIFTING IN ILLNESS

Richard said, "I don't know if you have ever had the misfortune of having the flu." I replied that I had. "Lying there in a sweat and you're drifting. And then somebody opens the door to your room with a click and your whole body shakes. You then feel like your soul takes a step sideways. In illness we get to a state where the realities begin to separate, because our consciousness is beginning to lose hold. The grip that the mind puts on the body begins to slip. A sudden crack like that has separated your soul shape. What we do deliberately and shamanically is lull the consciousness, shut the parrot up, put a bag over its head or put it back on its perch or something like that, and then crack! That shakes the body and then you are out. Other things we do are, to this day, technically feasible for a shaman to send a traveler . . ."

I interrupted. "Tell me about travelers. Are these out-of-the-body experiences?"

Richard explained that the first grade of an out-of-body experience is what the general public knows about out-of-body experiences. That is when you send your consciousness willingly out of its body and you experience things that you could not possibly have experienced otherwise. You see events, etc.

The second one is what is often called the doppelgänger in German. Bilocation is its modern name. In this case, the person appears to be in two places at the same time. Normally, the practitioner of that has no idea of what went on. What the shaman does is take those two techniques and marry them together so he has a projection that is physical, i.e., he can turn pages, press buttons, and other things. But the "double" is under the conscious and witnessed control of the shaman.

DREAM FLIGHTS OF THE INDIAN MEDICINE PEOPLE

Richard and Chris had confirmed my dream experience. It was possible to travel in this way by using the mind. In some sense you awaken in another body, which somehow is still your body. But what about journeying into the body of an animal? Could your soul find itself in a cloud or a rock? In the movie, Nexy certainly believed that evil spirits had inhabited the bodies of the pink dolphins in the Amazon River. I then remembered the meeting I had with the Chumash medicine people. They confirmed that all of this was possible. I still didn't know how, and I still wasn't convinced that these experiences were real, even though I had an experience of being in an animal's body in a dream.

I told them that I once had an experience where I went to sleep and when I woke up I was high up in a bird's nest on a rocky tor. I could see down from the nest and witnessed the ocean several hundred feet below me. I was situated toward the rear of the nest, close to the rock. Around me, I was physically able to touch with what I felt was my hand. I could feel the feathers of the other nestlings. When I say touch, I mean that these birds felt as real to me as the table feels real to me. I was in this nest with these birds. I could turn my head and see things. The colors were vivid.

I told them, "I have had other experiences where I enter into another world. It is not the world I went to sleep in. I am not in the same form sometimes. Is this familiar to you? Can you relate to this?"

Kote said, "Oh, yeah. This is a common experience for me."

"Can you do this at will?"

Kote said, "Oh, yeah. Sometimes I'll take off at night and A-lul'Koy will grab onto me and hug me and say don't leave me here."

A-lul'Koy said, "I'll go too. I used to stay. But once in a while I go with him."

"So you fly together. What would you do? When you are in that state, do you experience the physical reality of that state somewhat as I described my experience?"

They told me that sometimes they would have physical sensations. Most of the time there was no sensation at all. They would feel themselves passing through space. It would feel much like the air on your hair or face. Then all of a sudden they would "arrive." Then they would become aware of what is going on.

A-lul'Koy said, "I have noticed that if there is a purpose for us to be there, then we would arrive right at that point."

Kote explained that usually the everyday mind, if you take that with you, becomes a problem. So you must learn to leave the body and that frame of mind. Unless your "awareness" is really developed, you won't be aware of where you went or what you have done. It is just hit and miss.

Much of the time the reason for the Lotahs' flights is to do work. Sometimes it's going to sacred places and moving a rock an inch to adjust the spiritual balance of the land or the energy flow of the land. Other times it's working on people.

CHUMASH TELEKINESES

I wanted to know more about their abilities to move without their bodies.

Kote told me that when you start moving energy, you must regulate how quickly you travel. If you go too fast, you're liable to move into something and have a conflict. If you move too slow, something is liable to catch you from behind.

In other words, voyages into these other worlds were not always safe. My early experience with the "plane of suicides" concurred with this observation.

I tried to understand what the Lotahs were telling me in terms of quantum physics. Somehow, I reasoned, they were able to enter parallel worlds, other realities. But how they did this, I wasn't sure. I felt that they were possibly tuning to a movement in consciousness that was faster than light. I knew that quantum waves could move this way, but I had no idea if this was even close to what they were doing.

A-lul'Koy said, "You have to be careful. There is an old shaman that lives in Indiana. Sometimes we have to call him and find out if it is time for us to meet with him because of the things that we carry together.

"One time he came to us on a lightning bolt. There was a storm over the Rockies and an electrical imbalance. He went into the storm and he came out over here. He came to doctor a friend of ours. The family was in the house. And he was upside down walking on the ceiling, doctoring this lady.

"I have even had him come through my telephone a couple of times. I wanted him to look at a particular object. So I said, 'Howard, come on over.' You can feel the air get real dense as he comes through. He just builds up space. He'll look at the object and then he'll go back on the phone and he'll describe what he saw and we'll talk about it.

"Once I was talking to him and he said, 'A-lul'Koy, would you please go put some clothes on so we can finish this discussion.' I had just gotten out of the shower. He said it so cute."

I wondered if they could also transport themselves over a telephone line. They told me that they could but before they could they had to have a clear reason for doing it. Kote called it *intent*.

A-lul'Koy said, "It depends on the need. You can't just perform a show."

The Lotahs explained that there are a lot of things they would like to do, but there are traditional restrictions. Because of what they are, they, too, follow rules. There are some rules they will not break. For example, they will not go after people to afflict harm, unless they inflict it on them first. That is the law of nature—self-defense. It is possible to become evolved to a point of being so powerful that a shaman can just think somebody dead. It appeared to me that there must be some restraint; otherwise the shaman will be dealing with anger and vengeance, and that can backfire, opening the shaman to harm to himself.

SHAPE-SHIFTING AND ANIMISM

Now, here in the Lima cinema, the movie shaman Meliton is confronting the evil spirits called *yacurunas*—dolphin phantoms arising from the Amazon River. And as I watched his predicament, I remembered my own encounter with the dangers of the spirit world.

I had realized that out-of-body travel could be dangerous. It was becoming clear that these journeys were clearly more than wishful thinking or flights of imagination. They were, instead, flights that took the shaman to the edge of life and death. Thus great skill was required.

I then remembered stories about werewolves, and people changing form, and whether these stories had any basis in reality. This was a different phenomenon from, for example, my experience of taking on the form of an animal in a dream. So, the question was, if a shaman could travel out of body, at will, could the shaman also change his form at will and become an animal? There were a number of fictionalized stories about this. Chris Hall, the shamanic researcher I had met in Brighton, in 1988, a year ago, had told me that this was possible, and he related a story to me that fascinated me. We were driving through Brighton, shortly after he picked me up at my hotel.

THE BERSERKERS AND THE SHAPESHIFTERS

As the light turned green and we accelerated, I asked Chris about shapeshifters. He told me, as he turned a corner, that the ancient Nordics had warrior-sorcerers called *berserkers*.[5] A berserker, a name that means bear-shirt, was a "shapeshifter." Berserkers "turned" into bears. There are a number of accounts of people who "saw" these bears in battles. In one supposed eyewitness account, which was detailed in one of the "Icelandic" sagas,[6] a berserker went into his tent instead of fighting. This was quite strange, because a battle had ensued in which his Danes were terribly outnumbered. He decided that his best strategy was to enter his tent, go into a trance, and send out his bear-form to fight. This bear-form was literally a thought-form, not the real man in a bear suit.

When the opposition saw the bear, everyone ran away from it. If anybody tried to cut it down, the sword would pass through it because it wasn't really there. It was an illusion. This bear turned the battle momentarily in favor of his Danes. But, a little boy inadvertently awakened the berserker, and the bear vanished. Consequently, the Danes were beaten.

Chris told me that he had actually seen a shape-changing event. It was witnessed on a hill outside of Brighton, in February 1988. Someone who was heavily involved in neopaganism had taken LSD. He actually began to resemble a four-legged animal.

Other people, as well, saw this happen. And they all objectively agreed on what they saw. But Chris couldn't describe what kind of animal it was. It was to him a kind of mythical beast.

CAUGHT AS A RAVEN

The Lotahs (the Chumash medicine people) also discussed shape-changing with me. I asked if Kote, himself, had ever attempted to shape-change. Kote told me that A-lul'Koy caught him as a raven one day. Kote then laughed for the first time, spontaneously. A-lul'Koy said, "He scared the hell out of me. That was my first experience to overcome."

I asked what happened. Kote said, "I wanted to see what she was up to."

A-lul'Koy said, "Everyone else saw this raven that was staring at me

[5]According to Norse mythology a berserker was a fierce warrior who fought in battle with frenzied violence and fury.

[6]Taken from Ellis-Davidson, H.R. *The Gods and Myths of Northern Europe*. London, England: Penguin, 1977, pp. 66–69.

through the window. It was a real raven. We were in a car. He stopped himself in flight and was looking at me on my side of the window. The other people were saying, 'What is wrong with that bird?' They were freaking out and everything, and it was Kote. I saw Kote. They saw this raven."

Kote said, "I was loving it."

"And, of course, he wanted to get hold of me. So as soon as I got home I called him," A-lul'Koy added.

THE BIRDS OF DEATH

During the same time I was discussing the berserkers with Chris, he also told me that the old Nordic shamans could read omens from the flight of birds. By watching the flight of a bird, you can see, for example, that someone is going to get injured. Chris explained that he has a rapport with birds. Kestrels and magpies appear to him as messengers, telling him things. As Chris related this to me, I suddenly remembered the birds that appeared at the moment of my mother's death.

I knew she was very ill and that her death would be coming soon. My sister telephoned me at my California home to come back to Chicago, where my mother was resting at my sister's home. When I arrived, I went to my mother's room and found her lying in a coma. Even though it wasn't logical, I spoke to her and wished her well on her new journey.

That afternoon, I went shopping with my brother-in-law and my niece. As we were returning in the car, two birds suddenly appeared and began to fly directly in front of our car, perhaps only a dozen feet or so ahead of us and about ten feet or so from the ground. They continued leading us as we made our way to my sister's house. It was an uncanny message. I knew that just as the birds appeared, my mother had passed on. I mentioned this to my family. We arrived at the house with the birds having led the way the entire time we drove. I immediately ran upstairs and my sister told me that mom had died a few minutes ago. Although my sister confirmed it, the birds had told me just when it happened.

THE ANIMAL GUARDIANS

As these thoughts about my mother's death began to fade, I tuned in to Chris's voice. He was saying, "I began to see an absolutely huge number of magpies. If you come up on the high grounds above Brighton with me you'll be surrounded by the bloody things. Because I've accepted them, now there is actually an attraction between us. It is really a most wonderful phenomenon and experience."

Chris explained that these creatures are our guardians. He said, "If you spend any time in nature, you will find your animal guardian. We've all got guardians. It is some animal that is on the same kind of frequency spectrum as you. It fits in the same bit of the jigsaw somewhere. Whatever orchestrates the whole thing gets to you through them."

His friend and fellow shaman, Richard Dufton, sees foxes whenever anything of import needs to be communicated to him that he is missing by direct means. This usually happens when he is thinking too much and not getting any intuitive flashes. At such times, he sees an inordinate number of foxes.

He explained that shamans the world over, all have totem beasts. (You'll recall my earlier discussion of totems in Chapters 3 and 4.) He said, "They pick you. They are preordained. You don't think, 'I would really like to have a wolf because I think that it would be really keen.' It doesn't happen that way. It has to be an animal that somehow has a connection with you. There is no way to rationalize it. If it is a day-to-day thing for me, it will be a magpie. If something really heavy is going to happen it will be a kestrel.

"Other animals that I seem to have an empathy with are large cats. Tigers particularly. They always come to the bar at the zoo and purr whenever I am around. I have no idea why. Since I learned to relax with that I feel happy about it rather than think that this is strange. This is wyrd."

Chris laughed at the double entendre. It was wyrd and weird. He continued, "Sometimes I'm feeling like I'm going crazy. This is the problem. You observe these things. They are experiential reality. They have validity to you, but not in the parceled-out universe which has been assembled from leftovers from the scientist's table."

SENDING SPIRITS TO GUARD THE LAND

When we had later joined Richard at his home, Chris continued to tell me about using spirits as guardians. In old England, tribes, in accord with the American Indians, would send spirits to guard their land. Suppose there was an important boundary between two tribes. Two tribal kings, with the best intentions in the world, agree to have peace. But suppose a few wild young bucks from one tribe want to go over and rape a few girls in the other tribe. That's certainly going to upset a few people and probably cause a war. So to prevent this from happening, both tribes would take their oldest, wisest, and bravest warriors that had passed on and bury them all along the boundary. Those spirits would keep that boundary strip inviolate.

Richard told me that there is another level of spiritual existence. Not all spirits are of human or animal form. The earth produces levels of existence that the average adult never detects. But often this level is felt in childhood. This ability to sense earth spirits is usually lost because of parental influence.

As Richard explained it, "You take a chair to a child. Why can't it be a spaceship or a castle? It's only mommie saying, 'No, no, no, that is a chair. You sit on chairs,' that creates this reality. Children have the ability to cross over to other realities up to about five years of age, and then they get told off for playing with their invisible friends and things like that, and they lose this ability completely at about fourteen."

RECOVERING OUR LOST SENSES

It is always difficult to realize that we may have lost an ability that we possessed. It is known that the human fetus actually has rudimentary gills for a period of time while in the uterus. Even a tail is present at the earliest stages. But then, as the fetus evolves, these are lost, never to appear again. Surely nature intended that these other organs would be used; otherwise they wouldn't have appeared. Probably they were vestiges from an earlier life-form from which the human being developed.

But what about our ability to shape-shift as Chris explained it? Or to sense the spiritual level of a planet? Have we also evolved in such a way that we have lost these abilities too? Certainly my own childhood was filled with imaginary events. Could some of them have been the sensing of other worlds? Had I lost this ability simply because I was brought up in a Western household?

I was beginning to feel that we had lost these abilities. I wondered if it was possible to rekindle them. My inner knowing told me that these other senses, the five senses of imagination that went beyond the immediate senses we all know about, were still present. At least they were potentially present. It was somewhat like getting out of shape. If you fail to exercise for a long period, it is always difficult to get back into shape again. If you have not read a book in a long time, reading is tiring at first. Anything not practiced regularly will be difficult when tried again.

Perhaps the same is true for spiritual sensing. We can regain our other senses if we practice, going slowly and not getting too discouraged if something magical doesn't take place right away.

It was apparent that shamans knew they had these other senses of imagination developed. We, in the Western world, not believing in the reality of the imagination, dismissed all of this to the world of fantasy. They didn't. The imaginal realm was as real to them as what we call the physical realm. I then realized that we, in the West, are now faced with two basic problems.

The first problem is our failure to believe in any worlds of experience outside of what we call reality. The first step in rekindling our lost senses is believing that they exist. You can't see what you don't believe is real. Once you believe that your lost senses still exist, you need to become aware of them and learn to develop them.

The second is that because we fail to believe in the imaginal realm, we also fail to recognize information from it when it appears to us. Like our prenatal gills, we have given up half of our human senses, and in so doing we have become less human.

Now the only way possible to access these extraordinary realities is to take the function of concentration off the major realities—the so-called survival experiences that we call normal living—and put it on these other worlds. This is equivalent to shifting the focus of our attention. We defocus the background and refocus on the foreground. In other words, we must alter the dice game of probability. Right now, the dice are loaded so that we pay chief attention to our survival, even though nothing is threatening it.

This change in focus can be done only by changing our consciousness, our way of perceiving ourselves. We must learn to pay attention to that which we normally may not even believe exists.

This is not easy to do. We have an investment in ourselves. This we call our egos. When a situation develops in the world that we are uncomfortable about, our egos, much like the surfaces of our unconscious minds, freeze over. The bubbles of insecurity that came into the preconscious region between conscious and the unconscious form the structure of that surface. And we consciously operate in lock-step, in now-aware behavioral patterns. We march to the drum; we habitually follow the leader wherever he takes us.

So how do we break the habit? How do we learn ways to believe that alternate realities really do exist? A better understanding of such things as quantum physics may help in this regard. Accordingly, it is precisely how we observe that creates the reality we perceive. Change the "how" of it and you change the "what" of it. In other words, change how you see and think about yourself and you change what is actually present in the world.

Remember, according to the observer effect, what you perceive not only affects yourself, it affects the object of your perception. Both you and it are affected by the action of perception—by what you believe is "out there."

If you take a substance like ayahuasca or any other mind-altering substance, it will affect your perception of reality by actually, in a subtle manner, changing the reality you are in. Thus the reality is altered by your altered perception of it.

However, one didn't necessarily need mind-altering substances to alter one's perception of reality. The Chumash had been able to enhance their sensing abilities by using animals that had heightened senses of smell. They had become for a moment the animals that they were evoking. They were using animal fur in much the same way that a scientist uses a microscope to see the microworld of a cell. The animal fur connected the shaman to the animal. The shaman's consciousness and the animal's became one.

In every case that the Chumash medicine people told me about, I kept coming back to that one realization. Consciousness is a field that is not permanently held in our bodies. Consequently we are connected to all life, and that means that we have the ability to sense as the animals around us sense, even to sense beyond the walls of space and time.

HOW TO ENTER A PARALLEL WORLD

Okay. You believe parallel worlds exist. Now what? This is probably the most difficult step to take—learning to become aware of parallel realities when they present themselves to you. For example, out-of-body experiences occur all of the time, although we are not usually aware of them. When we sleep, we often journey to other worlds. But when we awaken, we usually forget the experience.

The shamans have been able to reach a state of consciousness that is like sleep, only they remain awake. They have shifted their perception. And more important, they recognize that their consciousness is not confined to their own bodies. They see consciousness in everything. By choosing to observe themselves as spirits traveling over telephone wires or floating in thunderclouds, they are tuning to the consciousness in these objects. This is really an extended self-observation and that is precisely what we must do to regain our lost senses and experience parallel realities as shamans do.

Let me explain. Everything in nature undergoes self-observation. It is a process wherein each observer defines that which is outside of his or her or its *self*. The key here is learning to extend what one calls one's self beyond the boundaries that we normally set up. Again this is not easy to do, because we are already so occupied with our own self-images.

To understand this better, consider that self-observation occurs even in atoms. The atom exists in stable energy patterns called "states." In order to maintain a state of energy, the atom must "recognize" itself, observe itself to be in a state.

Just how an atom can observe itself is perhaps difficult to comprehend. But if we, for the moment, consider that an observation is an interaction involving a transition between two states, a so-called observer state and an observed state, then it isn't too difficult to understand what I mean. When you observe something in the outside world, according to quantum physics, the state of the object suddenly takes on a discrete value. It appears in the world. And simultaneously, you become aware of it. Consequently your observing apparatus, whatever that consists of, jumps into a discrete state corresponding to the state of the object. There is an interaction between the observing apparatus and the observed object, and that constitutes the act of observation.

The same holds true for the atom. The atom continually "checks" its energy state by constantly involving itself in its environment, even if the environment is just empty space. Actually the electrons in the atom are constantly "dancing" with the photons of light they emit. So they are continually emitting and absorbing or, if you will, "observing" themselves.

If the atom alters the way it looks at itself, if it suddenly becomes aware of an electron's position, for example, the atom will change its state. It will emit a photon and it will undergo a shape change. It normally will do this if it is in an excited energy state.

The observer's consciousness of an atom is no different from the atom's consciousness of itself. It doesn't matter who or what the observer *is*. As long as it takes on a state corresponding to the observed object, an act of observation has occurred. Thus, I speculate, when we observe anything at all, there is a connection between our consciousness and the object's. For a brief moment the observer and the observed become one. Who the observer is in all of this, and where that observer is, no longer have any significance. Consciousness, like drops of water that coalesce, becomes one.

But soon enough, the observation ends, and consciousness once again perceives itself as separated from all other consciousness. It once again nestles itself neatly inside of the bodies of the observer and the observed. Every act of observation is a joining of consciousness into one greater consciousness. It is the forming of a larger "self" from smaller selves—the observer and the observed.

My voyages to the spirit world were reminders that our consciousness did not end outside of our bodies and that it was possible to extend it, perhaps to the ends of the universe and to the beginning and ending of time. Let me explain this more fully.

According to my understanding of quantum physics, all things, including our bodies, persist as they appear because they are evoking a repetitive pattern of self-observation. They continue to see themselves as they were. This pattern exists throughout the past, present, and future of the body. It actually goes beyond linear time. This pattern, which shapes the forces that keep the body as it is, is controlled by light interactions taking place within the body. All matter is continually absorbing and giving off light. When an atom emits a quantum of light, that light manifests as a wave. When it is absorbed by another atom, it suddenly changes from a wave into a particle of light. This wave-particle dance of light within the body is truly magical. Waves that suddenly change into particles and back again into waves actually give physical form and shape to the body in which they exist.

So who or what determines the shape and size of beings? The world of matter has limitations. The charge of an electron is a fixed number, and so is the mass of a single proton. That can't be altered. Consequently, a man cannot be physically as big as a building or as small as a fly. Even though the forces that hold us together are conducted by electromagnetic waves of light that transform suddenly into particles of energies, I can't see how we could ever gain that much control over matter, to violate the fundamental constants of nature. But we can learn to change our consciousness.

The field of consciousness is following a pattern, one that was built up over countless periods of physical time. These patterns are evolving, changing all of the time. And we, who deal with ordinary waking consciousness, are only realizing a small part of that pattern in our everyday lives. So because of our restricted self-images, we have limited access to the magic of our own bodies.

If we could become aware of our light-bodies (the light within our bodies), perhaps we would be able to shape-shift, change into animals, or fly across telephone wires, too. The key *is* in our consciousness. Even if we could not physically change our bodies, we can change our minds.

9

Time Traveling and Visions of Past and Future

In the film, Nexy voyaged back in time through her unconscious and faced her past life problems. In her shamanic journey, she watched herself when she was a child, as if she were another person, a visitor from the future. She saw how her past life was affecting her present, and she realized through the ayahuasca experience that she was on a quest for truth—her own meaningful life. She also realized that through her time-travel experience she was able to heal herself.

Later that evening I went over my thoughts. There was some connection between time traveling, the unconscious mind, and healing. But what was it? I let my mind wander.

Nexy's voyages into the past reminded me of my own interests in time traveling. Perhaps there was a connection between time traveling and the mythic and visionary world of the shaman. I recalled my second hypothesis: Shamans see the world in terms of myths and visions that at first seemed contrary to the laws of physics. Okay, but what about healing? When they healed anyone, they certainly were in an altered state of consciousness.

My third hypothesis had also now emerged: Shamans perceive reality in a state of altered consciousness. This altered state appeared to me to be a greater sense of self brought on by learning to observe oneself in all things, as part of a greater consciousness. Perhaps that altered state was connected to the mythic world, and that in turn was connected to time traveling.

It was clear to me that time traveling was a very important part of the shamanic journey. I hadn't read about this facet of shamanhood before. My reconstruction of my own quest had led to this surprising turn. It was simply a matter of logic, as far as I was concerned.

I put together the logical train of my thoughts. I knew that shamans use sounds and sacred plants in their initiation rituals. Thus my first hypothesis, that all shamans see the universe as made from vibrations, and my seventh hypothesis, that all shamans work with a sense of a higher power, were correct. Through a vibrational entrainment, probably with sacred places on the planet and through use of sacred visionary plants, a shaman gained a sense of power and altered his or her consciousness. That power was connected to two distinct ways of seeing the world—or two modes of consciousness.

There was the masculine, logical, rational, conquering intellect which dealt with thinking and sensations. Then there was the feminine, mystical, spiritual, yielding consciousness which dealt with the intuitions and feeling. It was the recognition of this sexuality in each one of us that would lead to a whole or healed person and planet.

I felt I was treading on quaking ground, but perhaps when a person was fully awakened to both sides of himself or herself, he or she would be able to sense the presence of other realities. These would appear as sounds, images, and perhaps even other senses would be invoked such as touch or smell. But where did these extraordinary impressions originate?

I began to realize that to be healed a person sometimes simply had to get out of the way of any fixed belief system. It was a form of prayer, a quest to allow whatever power there was, wherever it came from, to come into the person and heal. When something is wrong or not whole, the only way I saw that a healing could occur was through a belief that it was possible.

No scientific theory today can explain how healing occurs. I was gradually becoming convinced that healing would occur only through the unconscious mind making voyages to other worlds, journeys of the soul that were beyond chronological time.

My sixth hypothesis was that shamans voyaged to other parallel worlds. From my discussions, I was convinced that shamans could do that. And yet, paradoxically, these conscious journeys to other realities could lead to death and rebirth. The trick was to stay alive and still make the journey. If it doesn't kill you, it will make you well.

And, as I said, if all of this was true, then these soul journeys took place beyond the grid of chronos, space-time. That meant, according to the new physics, that these were voyages beyond light-speed and therefore not subject to the time ordering that must occur for all matter confined to speeds less than that of light.

Quantum physics also indicated in its essential calculation of probabilities that a quantum wave of possibility had to travel both forward and backward in time. If this wave was the vehicle of the soul, then time travel would be connected with death and possibly near-death experience. It was also connected with birth. I then thought back to the movie.

JOURNEYS OF AN AYAHUASQUERO

As I watched the film, Jorge Gonzalez, the Peruvian shaman, now the actor, was back on the screen. Miguel, still looking for Nexy, returns to the bar where he and Jorge had met. Jorge decides to take Miguel dancing. I laughed watching Jorge dance so well. The image of Jorge the actor and Jorge the shaman were overlapping again.

When we first met, he had taught me that it was possible to journey to the imaginal realm. It was within my grasp. Since Jorge was also an academic, I wondered how he ever was able to make the leap from academia to the shamanic jungle, to awaken his mythic sense. I certainly wondered how he ever became a shaman. He was a professor of education, although his father was a shaman who later became a brujo, a black magician.

Judith had asked Mike, Jorge's translator and his shaman-initiate, "How did Jorge get to be a shaman?" Mike translated her question, and Jorge told us that he was studying prostitution in Peru for his doctoral thesis. He found that one of the characteristics of the prostitute's personality was that she had a very magical train of thought. A very high percentage of them were clients of shamans.

Jorge had accompanied many of them to visit shamans. One of those shamans was later to become Jorge's maestro, Don Solon. When the maestro saw that he had possibilities of being a shaman, he invited him to be his apprentice. He has been a shaman ever since that day eighteen years ago.

I asked Jorge if he had visions of other worlds.

He told me that he saw many worlds, an infinite number. He confirmed my ideas about parallel universes as they are envisioned in quantum physics. When a person is moving through life and makes a choice to do one thing rather than another, the world that he or she didn't choose still existed. Thus it was possible that a person could be living many lives at the same time. He also confirmed that the future, the present, and the past all existed at once.

He told me that we are alive today because we are also dead tomorrow. We are today because we have been yesterday and we are going to be tomorrow, and that tomorrow exists today.

Jorge explained that it was possible to make an instantaneous movement from here-now to an entirely different location on the planet. One needed concentration, and if a number of shamans sat together and all focused on the same thing, it would be very possible.

I wondered if we would physically be able to transport to another world. Jorge then explained that he and other shamans traveled and lived in the world of a famous shaman, Maestro Delgado, when he was already dead twenty years. (Meliton, the movie shaman, was the son of Maestro Delgado.) They touched him and they talked to him. But with Jorge's father, who also had died, not only they saw him and felt him, but other patients also experienced him. The participants saw Jorge's father, and three of them saw the maestro Don Solon when he wasn't present. Those declarations were not made during ayahuasca sessions, but on the following day.

"When this was happening, you were also present here in the United States?"

"Yes."

"And you had a vision of this other reality. Is that true?"

Mike (Jorge's translator) answered: "It is my understanding that they came and visited with us."

"They came here, you didn't go there?" I asked.

Jorge replied, "They came because we invoked them. So in my sessions I usually work with a couple of spirits other than the spirits of nature. I am very interested in the spirit of one that is dead and the spirit of one that is alive. My father has been dead now for many years, and Don Solon lives in Iquitos."

I then went back to my original question. "What about going there? What you have talked about is something coming to us. What about us going to them and living in their world?"

"Yes. That too. We not only heal people that are here but can also heal people that are somewhere else. For example, I healed a little girl who had a cloud in her eye, a cataract. She was three hours distance from Tarapoto. About midnight this little girl woke up and told her mother, 'Mama there is someone who is pulling my eyelashes.' The father also woke up, and they all felt that there was a presence there. My brother, who is a Western doctor, was a witness before and after this. He saw the girl when she had the white cloud in her eye and after it was all gone. What's left is just a little scar."

"What did you see when this healing was going on?"

"I saw that the little girl had this cloud and I went there."

"That was like another world to you. You were here and you went there with your mind or what?"

"Both body and mind."

"Did you physically get up and go there. Did you walk there? Or was it suddenly you were there?"

"All of a sudden you are there. This is a capacity of the mind that we still don't know about. Even I cannot explain it to myself. There are things we do and we don't have a concrete explanation of how we do it."

"How did you prepare yourself to do this?"

"It was with a very special diet and a very profound exercise of concentration and then I empowered myself with our herbs."

Jorge's description certainly fit James Jeans's philosophy. If we for a moment imagine that this hard and sometimes cruel world is a dream and that it has no more reality than any dream, then moving from one world to another is nothing more than waking from one dream and participating in another. Our world certainly appears to us to be much more than a dream. Why? Perhaps, it has something to do with all of us. Perhaps this world of seemingly solid matter is a collective dream which all of us have given up the ability to totally awaken from. Oh, I know that we have built up a very rational way of seeing that, over the centuries of experience, has led us to see the world of matter as objectively real. No one of us walks through walls or is not hurt if hit by a speeding car.

But I still felt, in spite of this hard evidence, that everything in this world is made that way to convince us that there is a real hard reality. I was treading on thin ice, especially for a physicist. I am sure that this viewpoint would not be popular to those of us who believe only in the rational world. Something was hinting that inert matter was only an appearance, an illusion. There really was no solid matter. But how could I prove that?

I somehow felt that I was walking in a cave before the invention of light. I had walked in this cave many times. My forebears, who were without light, had walked in it and had taught all of us throughout our history just what was there and what wasn't real. But as I walked, I had felt something that told me that there was more to the world than this cave. I had to find a way to get out before I could test my intuition. Since there was no light in the cave, my eyes were never used. They had retreated into vestiges.

Even though, at times, a crack in the wall of the cave would appear and light from the outside world would shine in, my eyes, having not seen before, would register the phenomenon as imaginary.

My cave was time. My "eyes" that could see into time had degenerated over countless eons. I had lost the ability to escape from the cave of time. Or had I? Perhaps those glimpses of other realities that I had experienced were momentary awakenings of my "time-eyes." With my "time-eyes" developed, perhaps I could journey through time as easily as I journey through space. And if shamans were able to do this, they could change matter as easily as we can change what happens in a dream.

To convey this in a physical sense, if time did not exist, and in some sense time was an invention of our own minds, matter could not appear with inert solid properties. Inertia is resistance to change in time. Without time, there would be nothing to resist. Matter would appear as ghosts.

I admit that this idea was sounding crazy. But maybe we did have the inherent ability to time-travel as if we were moving in a dream. What was there in us that was vestigial of this ability? Was I imagining all of this, just daydreaming? Or was my inner core self trying to tell me something? Was the secret to time traveling, voyaging to other worlds, contained within our imagination?

This thought also fit in with the idea (discussed in Chapter 8) that we have lost the ability to travel to other worlds because we have lost five of our ten senses. I wondered would we also be able to see through the veils of time if we regained those missing five senses of imagination?

We really needed a new view of time. I felt that the traditional or shamanic view of time being entirely cyclical would help. This view is a nonlinear sense of time. Newtonian physics seems to be linear. But, in the cyclical sense, quantum physics is similar to the traditional world view. The major symbolic item of almost every shamanic tribe, or every shaman that I met with, and this now included a wide range from the British Isles to Peru, to Mexico, to the United States, is the circle. The circle is a symbol that represents a periodicity or a vibration or a movement or something that comes back to itself in harmony over and over again, while the linear Newtonian mechanical world view is the straight line, going from one infinity to the other.

These are two very disparate symbols. They represent two very strongly different viewpoints. The quantum physics viewpoint may be much closer to the circle than it is to the straight line. It posits that everything is produced by a double action. It is a movement that goes from a given state to another state and then back to the original state again. It is a kind of circular motion.

We call this the multiplication of the quantum wave by its complex conjugate. It is nothing more than a vibratory movement going from the present into the future and then coming back from the future to the present and multiplying together, much as two waves multiply themselves when a radio wave signal modulates a carrier wave frequency. This multiplication is necessary in order to compute the probabilities that quantum physics predicts. However, the image of the circular movement of the wave is important.

Thus, in a sense, we are living on that circle. The circle changes with our perceptions of the past and of the future. What we believe is going to happen affects how we live our lives now. If we think that there is no consciousness after death and that this life is all that there is, we will do everything we can to live forever. If we think that there is no tomorrow, we will live today as if the future did not exist. If we think that the primal or shamanic world view is important, we will change the way we live now in this complex world.

And I had certainly been changing my way of thinking about the world. Just the idea that my mind may be a time machine excited me. But to use this discovery, I had to somehow become aware of my unconscious mind, as Nexy had to.

As I considered Nexy's plight, I remembered the first time I had ever considered that time travel was a possibility using the mind. It was in 1972 when I was teaching physics at San Diego State University.

I had been teaching the course "Physics for Poets," and in the hope of stirring up interest in the ideas of the new physics, I did everything I could imaginatively think of to make the course interesting to the nonscientists attending. It was then that I had a brainstorm. I looked through the catalogue of films at SDSU and noticed that there were quite a few films dealing with science fiction themes. I wondered, could I excite students about physics if I showed them these films?

I remember showing one film that certainly excited me. It was called *La Jetée*, and it was made by an American film-maker, living in Paris, named Chris Marker. The film was shot by using a series of still frames following each other in a dissolve format. Thus one frame dissolved into the next. Each consecutive frame was just a slight advance in time from the previous frame. This gave the film an eerie ghostlike quality and was very effective.

The subject of the film was time travel. It seemed that in the near future we will have nearly destroyed our planet. A society of scientists and psychologists living in that time, having perfected the theory for time travel, decide to try to send someone back in time to the period just before the disaster reached epic proportions. Perhaps the time traveler could perform some action that would avert the impending disaster. Only to do this they must find someone who has a fixation, a predilection, for that particular time period.

Confined in political prisons are certain reactionaries that had predicted the disaster. One of them is psychologically tested and found to have a memory of a particularly beautiful and sensuous woman who lived in the past—just about the same time that the scientists want to return to.

So the scientists and psychologists take him to their laboratory and begin their preparations. To time-travel is, it turns out, simple. The traveler simply has to go into a drug-induced trancelike sleep. The psychologists realize that if the dreamer has a fixation about any past time period, they can control his dream and to the dreamer it will seem as if he were actually back in time. In fact, the time-travel theory predicts that if you can dream the dream with great accuracy, you will forget that you are dreaming and all of your senses will tell you that you *are* back in time. In other words, the dreamer's dream becomes, for him, total reality. (Much later, after my lucid dream experience, I realized that the movie dreamer's experience resonated with my own lucid dream described in Chapter 8.)

So the experimenters send the traveler back in time through his drug-induced and controlled dream, and the dreamer realizes himself back in time. The experiment succeeds and the impending disaster is halted, but not without the unexpected surprise ending that involves the dreamer in more than he could have bargained for, his own death.

The film left me with a magical feeling. The technique of using single-dream dissolves coupled with the time-travel theme had a hypnotic effect on me similar to that of my earlier experience in Nepal. It altered my sense of time. It also got the students thinking about time in a way that no course in physics could have.

It was then that I began to realize that my quest was going beyond the traditional path of a physicist. Somehow the mind of the observer became more interesting to me than the objects of the observation.

It was becoming apparent to me that time was a prison—or, as I've mentioned, a cave—and if we could break out of it we could really see what the universe was doing to itself.

It was probably this visionary idea that led me to believe that some form of time travel, at least the ability to move consciousness to other time realities, was possible. Later, after completing my book *Parallel Universes*,

I was convinced that, with the existence of parallel worlds, it was indeed possible. But where was I to find evidence that time travelers actually existed? I then realized that shamans, throughout our history, were capable of at least seeing into the future and the past.

It was funny that it was a movie about time travel that in a way led to my being here in Lima watching a movie about another time/mind traveler. My mind returned to the shaman film once again. I looked around the theater. I had the uncanny feeling that by watching the movie I was also seeing a glimpse into my own future and my own past. I knew that near-death experience changed people's health and mental peace radically. But not everyone would want to go through that. What about time travel induced by shamanic trance? Perhaps such a technique released people from unconscious limitations. I remembered all of the shamans I had talked to about time travel and how it releases you from unconscious limitations that you no longer desire. I first remembered Richard Dufton and Chris Hall in Brighton.

Chris told me that one way the early Celts and Saxons had of seeing the future was with a technique they called *scrie*. We get our word "describe" from this word. To see into the future, they looked into a bowl, called the scrieing bowl, that was filled with water and hanging from three chains. In the bottom of that bowl lay a little model of a carp or a salmon. This was the salmon of knowledge in Celtic and Germanic literature.

Richard then said, "The Celtic shamans used to look into these bowls. Obviously the scrieing bowl used as a magical device is quite rare. The number of these that survive into modern times is even fewer still. But I do have the fish from the bottom of one from the Bronze Age."[1]

I looked at it with difficulty. I felt it had a power to induce a trance state if stared at for too long a time. I couldn't stare at it very long at all.

Later, when I visited with the ley hunters, Paul and Charla Devereux, in Wales, they told me that they had direct experience with visions of other times. We were meeting at the Wellington Hotel in Brecon.

TIME-SLIPS AND SACRED PLACES

Paul told me that mysterious occurrences took place near one of the prehistoric stone circles they had looked at in Oxfordshire. It had a little country

[1]Richard had it reliably dated at c. 800 B.C.

road going by it. They had been studying background energies at sacred sites, and had found a thousand-foot stretch of road with very high radioactivity. They first thought they had discovered a uranium deposit under the site. This would have been expected, because, as I mentioned earlier, a lot of sacred sites are close to uranium and other mineral deposits. In fact, that's what a lot of the land-rights issues of the Australian aborigines and American Indians are all about—they have sacred sites on areas that the governments want to mine for their uranium deposits.

But then they saw that there was some packing under the road that appeared to contain radioactive rock. That was not the mystery, although what happened took place along this same patch of road. It was a remote place where people had been studying the Dragon Project (see Chapter 5). Paul had evidence, provided by reliable people, of three cases—all taking place within a short period of time on that specific radioactive stretch of this road—that some form of paranormal phenomena occurred.

One man saw a car with two people just disappear from the road. Another person saw a gypsy caravan on the same stretch. It, too, appeared and disappeared. A third person, a microbiologist, saw a very big hairy animal for a brief second that also disappeared. They all reported their experiences independently. So it wasn't that one of them was affecting any of the others. One of them still doesn't know the others.

Paul believes that the experiences were essentially time-slips. The car was going to come, or had been there the day before, or week before or whatever. The same is true for the gypsy caravan.

Charla remarked, "It was known to be an old route where caravans could have gone in that particular area. That was something that we learned later on."

Paul explained that other researchers, whom they had consulted for his new book, told them that high levels of gamma radiation produce a special, localized kind of light phenomenon. I certainly knew that this was true and that such radiation was found in water-cooled nuclear reactors. Paul believes that there is a relationship between patches of high background radiation, time-slips, UFO phenomena, and this special form of light phenomenon.

Paul told me that he had read some of my own work about electrons going backward in time and other strange things in the new physics. He wondered if some macroscopic transient subatomic effect was occurring in these patches. I didn't think that this was possible.

By this time the lunch crowd at the hotel had taken over. So the three of us went to their house in Brecon, which also serves as the Center for Earth Mysteries Studies.

THE RAILS OF TIME

But could it be possible that in those times, when there weren't roads and cars and other Newtonian machines, people were able to simply sense alternate realities using sacred sites? Were they able to achieve a spiritual state of consciousness along a ley line? Did they walk these lines? Did they lie down on them? What did they actually do with the ley lines?

They must have had a reason, based on their own experiences, for keeping these sites sacred. I reasoned that even before the discovery of the scientific method, people were still eminently practical. They used these lines in some way. I had speculated earlier that at sites where high background radiation occurred, certain people might have been highly sensitive. This sensitivity could have induced them into what Michael Harner called the ssc, shamanic state of consciousness. Paul and I both suspected that the mind shifts that occurred there might also have allowed time-slips to occur.

The time-slips that occurred in Oxfordshire did not take place on a ley line. They occurred next to a site, a stone circle. Paul and Charla discovered it because they were looking at the site and its surroundings in their investigations of ley lines. They had made lots of Geiger counter readings. In the course of several years, they had built a background radiation profile of the site and the area around it. Apart from the rise and fall of radiation within the rings of stone, the radioactivity on that road, which ran along the ridge the circle was on, was three or four times the background count. It was over a specific three-week period that Paul received the reports of time-slip sightings.

The stretch where these events occurred and where the background radiation is quite strong, around 30 feet wide and 400 yards long, is a fairly flat terrain at the top of a ridge. There is nothing weird about its appearance. Paul said, "Interestingly, light phenomena, 'ghost lights,' tend to haunt roads, particularly in the United States. Is it some factor in the roads? Maybe instead of aimlessly wandering around, they run along these things like on rails."

TIME TRAVELING TO THE PAST

In the film *Slaughterhouse Five,* the hero finds himself slipping through time, seemingly out of control. My mind was doing this now while I watched Miguel in the movie attempting to make sense of his experiences. He returns to Meliton to find out more about himself and why he had been so captivated by Nexy. On ayahuasca, he learns the reason as his mind slips through time to his earlier experiences in Lima.

Was it possible for someone who was not a shaman to experience a time-slip? Apparently the ley line researchers had verified that it was possible. I then recalled what I had learned from Candace Lienhart, the American shamanka, when I visited her last year in Arizona.

She had also told me that she time-traveled. Her technique was based on being able to enter into a form of trance state, reminiscent of the movie *La Jetée*. She explained that anyone could use her technique with some training. A person would use time traveling to solve a problem that had its roots or its origins in an ancestral mode, some particular situation in life that is now bothering the person. She explained that what one has to do is set one's ego and identity aside and then say to one's self, the knowing self, not the ego or identity, that one wishes to travel back in time to the origin of where and when this particular situation or problem arose.

One states that one is looking for completion at some point in time. When one makes intention to do that, immediately the voices of the ego and the identity will start causing a lot of noise. For example, they may raise your temperature. They may give you doubts, saying that it is impossible and you can't do this. To mollify this, you must keep reassuring your ego and identity that you are not going to threaten or hurt them. They just have to sit quietly while you go back in time and do this task. You reassure them until they understand.

Candace said, "You don't have to keep reassuring them and attend to them the whole time. You set them in place and then you go. They may be noisy alongside of you, but you can't allow your mind to be distracted."

SETTING YOUR MIND SPINNING

I imagined that I would be lying down and relaxing or sitting up in a meditation posture, thinking about a problem that I had. I would then ask myself, as if I were asking another entity, to show me back in time where this problem arose. Doing this, she explained, I would begin to feel as if I were entering a vortex, a spinning of my mind.

She told me that the vortex adjusts its own speed, and it adjusts your "density," which causes your thoughts to start shifting. Again, the idea of density meant the extent to which you identify your consciousness with your body. By attempting to time-travel, you were essentially freeing your consciousness from its immediate body concerns. The more concerned you were with body survival, the less likely you would be able to do this.

As I mentioned, your identity and your ego normally can't handle this.

In such a state of consciousness, they're left out. You must move on beyond them and then you are able to move through time. It is here that you must have a clear intention to travel back in time to find the point of origin of a situation that you need to complete. And when you do this, when you arrive at your destination, you will engage every person, place, and thing that was involved in that conception, that moment of beginning, whether it's a thought or just an impulse behind the thought.

Sometimes we have unpleasant thoughts about the past. When they occur it is an opportunity to go back in time and correct a situation.

I asked, "When you go there, your intention is to complete what is incomplete. Is that enough of an intention, to say I want to complete what is incomplete?"

Candace explained that this wasn't enough. You need to have more in your daily life showing you what kind of completion you are attempting. One looks at one's problem in today's life. What are the issues of this problem? She said, "Let's go before that issue was decided. Let's go back before that manifestation arrived, to the thought that began the manifestation."

Once you arrive, you have to do an interview with the consciousness that began that manifestation. It's very important to not get locked up with the identity of that past consciousness. You need to keep your witness persona separate. If you identify with yourself, back in time, you will become that identity too, and then you won't really see what the intention behind that decision was. You must link up with that consciousness, but you maintain your separate identity.

You cannot always know what world you are moving into. You may not know the shapes and forms. You cannot count on any of that knowingness to go with you back in time. She said, "If you do get involved with human beings, things with shape and form, you may—and I have done this—find them turning around and looking at you and saying, 'Who are you and where did you come from?' "

HOW TO BE AN ANGEL IN THE PAST

I realized that when you time-travel, you may appear in almost an angelic form to those in the past you meet.

Candace agreed. When she time-travels, she tells the people she encounters that she is an angel and that she has a message for them from the future. She dialogues with her past. What she wants to accomplish is to provide her past with choices, alternatives that may not have been known to that past. She said, "If I could give them a bigger perspective, then it allows them free will to make a different choice. If you allow that choice, the present will change."

I knew how the present would change. We all have fixed in our deep unconsciousness memories of past events. These may not be the actual experiences we had but consist of memories and rationalizations of those experiences. Often the deepest parts of our unconscious mind contain traumatic experiences. Many times these traumas are contained within the tissues and cellular parts of our bodies. Muscle cells that become overly tense can be relieved if that tension was a pattern held in by a past trauma.

One time many years ago, I had noticed an older man who seemed to hold his body in what I thought was a very awkward posture. Later I asked him about his "macho" pose. He was quite small, and later I learned that he had been in prison for over fifteen years. He was unaware of his body posture. I soon reasoned that he "had" to walk that way to keep from getting beat up by other inmates, and even though he had been out of the "joint" for five years, the man was still in prison. It was the prison of his own memory now locked in muscle cells. Those memories had embedded themselves into his body armor. I could only wonder how he might have changed if he could have recalled the time when he began to hold his body that way.

By bringing memories of our younger selves into view, and then dialoguing with them, we can clear up problems we have now because of them.

Jorge had told me how this works. So had the shamanic researcher Dr. Jacques Mabit. The body is the unconscious mind. But when anything buried in the unconscious is made conscious, the body cells that contained that memory transform. Candace had made it clear to me that this transformation happened instantly.

Candace called the process "lobbying" the past to use its own free will to change. She said, "Usually the people in your past don't have full information, so when you give them full information from the future they are incredibly grateful. And when you release the foundation that all the misinformation of the past was built on, all the confusion, the nonclarity, every descendant of that line thanks you because they are released from a stress point in their consciousness and in their physiology, if they are still physical. It releases a thing in them. But again you've got to honor free choice. They don't have to change."

A UNIVERSAL PRINCIPLE OF SUPPORT

Candace explained that the universe is not constructed malevolently. She said that there is a universal principle that says that all people and all life in the whole fabric of the universe will support you. We may not recognize this, but it nevertheless clears a path for us and takes care of sideline details. When you time-travel, you become an ambassador to the past.

It's important to realize that when you go back to the past, especially when you visit your own past—say, a childhood incident—you go back to give free choice. You are essentially coming with information from the future to your past points, and then the past points are enlightened as if they suddenly had a vision of something new, something that was reassuring. At that point in time, they have different information upon which they can make choices. But you cannot interfere with it. They always have the option.

This meant that there had to be parallel pasts. Candace agreed and saw parallel universes back in time, as well. But you cannot attempt to force one universe over another, claiming one as more real than the other.

It appeared to me that all possible pasts could exist simultaneously. By visiting one of them, you were, in effect, altering the way that that past connected with this present. It was like constructing a new roadway between the possible pasts and this moment now. New roadways to the past meant a new way to envision the future. Perhaps this present was being visited by future selves. If this present is on a destructive path, these entities could be coming back in time to build a roadway toward them. The destructive future also exists. A roadway toward it from the present is already built, but no one is alive there, and thus no voices from that future are heard. I only hoped that some of us heard beyond the deathly silence.

Then, I remembered my surprising vision of another time-world, after taking ayahuasca for the first time. Although this vision occurred months before I was to come to Peru, this vision was of the ceremony that I had just seen a week ago in Cuzco. It was the Inti Rahmi. When I had the vision I had no idea that I would experience the same thing in Cuzco a few months later. Again I had the uncanny feeling that my vision had something to do with the connection between time, intuition, and shamanic healing.

The Inti Rahmi had begun. We were once again celebrating the beginning of the year and the return of the sun from its farthest journey to the north. It was winter. Now the sun would return to its people and give us its bounty. The drums and the horns signaled the beginning of the ceremony. The Sun God entered the vast sacrificial ground from the east corner of Sacsahuaman*—the Satisfied Falcon. Thousands of*

Incas, consisting of laborers, artisans, warriors, dancers, maidens, concubines, and priests, watched.

The Sun God, my father, rode on a high gold throne carried by ten slaves. He approached the central holy platform. The throne was lowered to the ground, and he walked through the hordes of worshippers, who stood with their heads bowed, and mounted the stone stairs to the high flat stone elevation. The long horns blew again. The drums tolled like bells. He turned and faced the sun, and everyone below sank to his knees and prostrated himself. The god in man's form held his hands to the sun and offered his prayer to his father.

My father moved to the altar at the end of the platform. He picked up a golden goblet filled with a white milky liquid, a pulque, and poured it into a large barrel-like container. The priests then rose to their feet and ascended the stairs to form a circle around the golden god, my father. The god poured the pulque into the priests' goblets. Then the people, still prostrate and surrounding the platform, all arose at once.

The maidens, each dressed in colorful flowing robes, mounted the platform. They reached into the vast container with vases and brought forth the pulque to distribute to the people. They descended to the ground and distributed the holy drink. Everyone drank the brew.

I waited in the ceremonial room below the altar. Eighteen Inti Rahmis had passed, marking my lifetime. Earlier, my father had come to my room in the royal chambers. He told me, "You now have a choice. You may be a priest and perform the sacrifice, or you may choose to be sacrificed. What do you wish to do?"

I knew without thinking. I said, "I want to be sacrificed. I want you to sacrifice me." I knew why I wanted this. I knew that I would return to my home in the sun. I knew that my father would bind me to the altar and cut open my chest, taking my heart and innards from my body. I knew that he would hold my organs to the sun and that all the priests would gather around my pulsating heart and read the signs for the next year. I prayed that I would be a worthy sacrifice and that the signs would be good.

My father looked at me with tears in his eyes. He said, "It will be as you wish."

The long horns sounded. It signaled the time of the winter solstice. People became happy. It meant that a royal sacrifice was about to be performed. I heard the laughing and the "hoo-hoo" sound of the Chasqui Runners as they circled the ceremonial platform, making everyone laugh at their antics.

I got up from my couch in the lower room. It was time. I ascended the inner stairs, the holy stairwell reserved for the spirit-being that would ascend the final stairway to the sun. I came up slowly, adjusting my eyes to the bright winter sun. I stood in the center of the platform. I wore a simple pure white robe. I was blinded for a moment. I was bathed in gold.

The priests bowed down to me. The maidens, with their hair long and braided, came up the stone steps to the platform. They approached me and formed a circle around me. Except for my mother, I had never even seen a naked woman before.

The maidens were beautiful. They all opened their robes, exposing their bodies to me. I was transfixed, and tears came to my eyes. Their beauty overwhelmed me. I felt faint. My father stepped through the maiden circle. I heard the priests chanting that soon I would be with the sun maidens. I felt wonderful and alive. My father took me by the hand and led me to the raised sacrificial altar. I lay down on it. A priest opened my tunic, exposing my chest to the sun. I felt the priests taking hold of my arms and legs. One of the maidens was holding my head. I looked up at them. Their faces were barely distinct in the bright sun. They all were smiling at me with tears welling in their eyes. I looked up at my father. He was holding the golden knife in his hand. I saw it strike downward. I felt the pain as it penetrated my chest. Then I became the sun.

Why was I a sacrifice before an Inca or Aztec priest? Was I really an Aztec or Inca prince, and this was a recall of a past life? Why was I asked, would I rather be the one who is sacrificed or the one who does the sacrificial act? My visions seemed to be a story, and I was somehow just following out my part as if I were an actor. I "knew" that I wanted to be sacrificed and that I couldn't cut out the heart of another. In the vision, I understood that it was an honor to be sacrificed and that it was an act performed by a father on a son. I didn't know whether this was true, but it seemed to be true, and it fit the biblical image of Abraham and Isaac.

Then when the vision faded, I saw all of the people I knew and had let down in some way. I saw the hurt in their eyes and knew that I had betrayed them and that they were now my enemies.

Aside from the psychological meaning of such an experience, the vision was a time-travel initiation for me. I was remembering a past life experience, or imagining that I was, and I was "remembering" a future life experience, my actual attendance at an Inti Rahmi ceremony just a week ago. And I was sitting here, in Lima, watching a movie. I had three time-lines running simultaneously in my mind. In some very real way I was being prepared to undergo time travel. It was at this time that I began to feel that illness had something to do with that ability. Then the "real" ceremony flashed back in my mind that occurred barely a week ago.

I remembered traveling by bus to Saqsaywaman, the fortress where the ceremony was to be held. The ancient city of Cuzco, when it was first laid out, had the shape of a puma, and at the head of the puma was the holy temple called Saqsaywaman. My tour guide jokingly told me that to pronounce the name, Saqsaywaman, you only had to say "sexy woman."

"Sexy-Woman" also serves as an amphitheater that could hold over 100,000 people standing and watching. Inca noblemen had their council house in this place. The last Inca emperors governed from this location. Astronomers used Saqsaywaman's watchtowers to sight the seasons through the stellar positions. In some sense, it is similar to the Acropolis of ancient Greece. Its stone structure is equally amazing. Each stone, some weighing over 100 tons, fit perfectly in its walls.

As I waited for the ceremony, I noticed that the crowd was becoming slightly restless waiting for the Inti Rahmi to begin. As my mind pondered, my reverie was broken by a sound I had heard before. It wasn't here in Peru, but in Nepal. It was the low brassy sound of the long horns. The Inti Rahmi was beginning.

I watched the whole ceremony in fascination. The details of the ceremony were much as I had seen in my vision. Only this time, I wasn't the one sacrificed. Instead of me, the sacrifice was performed on a llama. The priests cut out the llama's heart and innards and held them up for the crowd to see. There were black spots. That meant a bad year was coming for the Peruvian nation and for the llama herd.

It was quite a spectacle. At times I wondered if it was just that, an act carried out for the benefit of the spectators who had paid for their seats in advance. But, no. There was clearly more going on than that. It was too human. It lacked the Hollywood flair for accentuated drama.

But it confirmed that if time travel was possible, then surely it had to take place in the mind. My witnessing the ceremony and the vision confirmed this for me. But I was still doubtful. Perhaps it was just a coincidence, whatever that meant.

Then I remembered once again the first time I met the Eagle-man, Ed McGaa, just a few months before coming to Peru, while I was living in Santa Fe. Again it was another clue connecting time and shamanic healing.

EAGLE-MAN-OGLALA-SIOUX

Jamie Sams, the Seneca medicine woman whom I had met, had invited me to her home in Santa Fe to meet Ed, who was another Native American. She had told me that he was a shaman-initiate associated with the Oglala Sioux tribe in the plains of the Dakotas and had been given the name "Eagle-man."

I told him about my interest in the new physics. I wanted to know if he had anything in his background that would make it possible for him to time-travel.

Ed pointed out that it was important to be able to recognize the spiritual element in all things. I agreed. The new physics had certainly opened a door to this brand new vista.

To me, this new vision of physics had magic and had a clear mystical or shamanic component which cannot be ignored. I told him that, according to what the new physics indicated, it was possible to have visions of the future and the past, and it was also possible for an atom or subatomic particle to be in several places at once.

I then asked Ed if the shamans of his tribe were capable of that.

He told me that they were and that time travel was something that these Native American shamans did. I told him I wanted to learn how they did that. Little did I know at that time that he and his ancestors were checking me out, at that very moment, and that Ed was about to take me on a brief time-travel exercise of my own.

Ed asked me to join him in another room. I had told him about my mission to write this book. He reaffirmed that my mission was a noble one and that he would help me, providing I checked out with his ancestors. I wasn't quite sure what he meant by that. He then showed me an old holy object that he carried with him wherever he went. At first glance it was just a flat piece of rock with many layers and colors embedded in it. But then he said, "Can you see the Indian holy man's face on the edge?" I confessed that I couldn't. He said, "Look carefully at it." He pointed out the eye of a man's face shown in profile. I then began to see the profile of a face. The variegated layers surrounding the forehead resembled an Indian's feather bonnet.

He pointed to another face in the rock. I couldn't make this one out at all. He said it was his face. Then I pointed to an image that I could make out. It was an elk. It appeared very clear to me. He said, "This rock is very old, perhaps millions of years old. I found it in a stream on the Sioux reservation. When it appeared to me it shone as a light coming from below the surface of the stream. That light was as bright to me as if a flashlight had been submerged. This rock connects me to life that existed a million years ago. There is no time between then and now. And now this rock connects you to a million years ago because you saw in the rock something I never saw before. The elk. Thus I am millions of years old and so are you."

I failed to get his logic, but I realized a truth in what he had said. Certainly, I saw the elk's profile. He didn't. But once I pointed it out to him, he also saw it. This was a kind of observer effect. In a sense the elk's picture was not there until I saw it. In that sense I "created" the elk in a rock millions of years old. In a sense I had reached back through time and "made" the rock with that image in it.

How did I do that? Let me explain. According to quantum physics, if there is no observation made of an evolving system, then all possible scenarios must evolve together. For example, an unobserved atom moving along cannot occupy a single position in space until that position is actually observed. Until an observer "sees" the atom, the atom must occupy an infinite number of possible positions, simultaneously.

This simultaneous occupation of an infinite number of positions for a single system is known as the "parallel universes hypothesis." When the universe first began, over 15 billion years ago, no attempt was made to measure the radius of one possible evolving universe. Although each universe had a possible radius, since no one observed that radius, *the universe* at this time did not possess a well-defined radius. But then a mysterious interaction occurred and a radius was defined. How did this happen? If we ascribe an observer in all of this, then we raise the obvious question, who is the observer and when did the observation occur? The answer may be, as surprising as it sounds, we and now.

By peering back in time, by looking out into the universe at light signals that were emitted millions and millions, perhaps billions and billions of years ago, we may be the observers that are causing the early universe to become defined, and thereby we are choosing by our observations today what the radius and other physical parameters of the early universe were.

This is an example of what visionary physicist John A. Wheeler calls "delayed choice" measurements.[2] Accordingly, it is our choices made now in the present that determine what the past had to have been.

[2]See Hellmuth, T.; Zajonc, Arthur C.; and Walther, H. "Realizations of Delayed Choice Experiments," in *New Techniques and Ideas in Quantum Measurement Theory*, ed. D.M. Greenberger, Vol. 480. Annals of the New York Academy of Sciences, December 30, 1986. Also see Wheeler, John A. "The Mystery and the Message of the Quantum." Presentation at the *Joint Annual Meeting of the American Physical Society and the American Association of Physics Teachers*. Jan. 1984.

Perhaps this was true for the stone I held in my hand. And then again perhaps it was just a myth.

What did myth have to do with time travel? I was reliving some form of myth, now, in Lima. I was "seeing" not just the movie, but the movie of my life. I was witnessing myself in a parallel world. Time as I knew it was not making any sense. I had to struggle to focus on the film and not think of my own life. My mind shifted back to the Lotahs, the Chumash medicine people, I had met a few months earlier.

THE PAST AND THE PRESENT ARE THE PARENTS OF THE FUTURE

We then talked about seeing into the future and the past. Kote explained that when you look at the past and the present, you understand that they are the mother and the father of the future, which was similar to Candace Lienhart's assertion. When you know that, anybody can predict what is going to happen.

I pointed out that maybe it was possible that the future could come back into the present and talk to us.

Kote said that this wasn't possible. He said, "The experience isn't there—destiny and all that stuff like that. This is on a moment to moment level created by the individual."

I was somewhat amazed at Kote's viewpoint. I expected that since he believed that time was an illusion, he would have described another vision of the future. I thought that he would affirm that in some sense the future was already present.

I had my reasons. In the mechanical world view, the old world view espoused by classical physics, the idea was that if you could know the causes, you could predict the effects. So if you could say what happened in the past, with great accuracy, you could predict what will happen in the future. The laws of nature follow mechanical actions.

But this was shown to be in error with the discovery of quantum physics. In the new physics or quantum world view, it doesn't work this way at all. Instead we must deal with probabilities, not actualities. Nothing is absolutely predicted for the future, and nothing is actually determined as the past. Both the past and the future are connected to the present as possibilities.

But there was the parallel worlds interpretation of quantum physics to contend with. According to it, these possibilities, both in the past and in the future, were real. The possible pasts actually happened and the possible futures will actually happen. They only appear as probabilities because in any single world the other worlds aren't present. Thus they appear as possible worlds.

And there were also the least-action pathways that thread through these worlds. By changing one's thinking, it was possible to reorient one's path, creating a new least-action path which connected to these other worlds. To the person experiencing this reorientation, it would appear to be a single world. By changing the path, by becoming aware of other realities, one was constructing an individual reality.

Thus the so-called past really was a superposition of all the possible pasts for the individual. The elements of the superposition were separate realities. But for the person they would appear as possibilities or tendencies. What actually was the real past for the person was impossible to determine. That past could change if the person's memory changed.

When shamans journey to the past or the future in attempting to heal someone, they are only picking up the tendencies that exist in the mind of the patient. From a parallel worlds view, what actually occurred in the past is unknown, but is carried with the patient as memories of real events. But these events are continually being reconstructed by the person in the present to make sense of the present.

That meant when a person journeyed back in time to "the" past, she or he was doing no such thing. She was reconstructing that past as she remembered it.

This was an important insight into healing. Often what we recall as the past justifies our present. We may keep in our memories pictures of the past that really are keeping us ill. For example, the memory of a parent who wasn't nurturing can sometimes keep a person ill by reaffirming that the person isn't worth nurturing. A similar situation exists for the future, only there it seems more evident, since from the point of view of the present, the future has not happened yet.

As I was thinking about all of this, Kote said that they do time-travel when it is necessary. Usually they did this when the journey had to deal with understanding something on the emotional level. Either one of them or sometimes both together would time-travel. He then told me of a particular incident when it was very necessary. I could tell that this was a sensitive issue with the Lotahs. He said, "We had a tribal law that forbade medicine people from marrying each other. A-lul'Koy noticed this and told me, 'Well that isn't right, because medicine people come together and they produce medicine children.' "

A-lul'Koy continued the story, "It stopped me from continuing my tradition as a medicine woman. I wanted to experience another part of life, to give birth, and this law said no more. I knew that this wasn't right. My mother was a medicine woman. All these doctors were in my line, and now I was told that I must stop. I wouldn't accept that. I wanted to know why. So I went to Kote and he said, 'Let's find out where it came from.' "

Kote continued, "So we set her off lickety-split down the time-line. She went back about 409 years and found that a Jesuit priest had inserted that crap into our system. Over generations it became law."

"You found this out by going back through your own genetic memory cycle?" I said.

"Mine and his. I had to go through both," A-lul'Koy answered.

"And that information was there?" I said.

"Yes."

"When Kote put me through that 400-year journey," A-lul'Koy said, "I went back until I felt a disturbance in frequency. Then I couldn't go back any farther because something had stopped. At that point I saw what had happened. What stopped me or stopped the frequency was judgment. When I felt that vibration of judgment, I stopped and checked it out."

"Judgment is a voice coming from the past, isn't it?" I asked. Kote agreed. I continued, "So it itself can be opened to see who was the judge. Is that right?"

"Yes," Kote answered. "We almost started a tribal war. We were going to have a child. When A-lul'Koy was five months pregnant with this child, some of the tribal people said this wasn't going to happen. They, through their emotional outrage, killed the child. We're not ready yet to get on good terms with them. It's not that we're angry; it's that they are dangerous. They are Christian-thinking people with the powers of the past. When that anger is projected, it is strong. They have no understanding of who or what they are."

"I believe that we have the right to breed our kind. Just like the owls and every other animal have the right to breed their purity, we have the right to breed our kind," A-lul'Koy said.

"By breaking the law and going back to the old, old way we came into conflict with the system. Our type of people is always on the outskirts of everything," Kote said.

INTENT AND A BEAM OF LIGHT

Kote explained the process of time traveling as he saw it. The simplicity of what goes on confuses everybody. You simply have the intent of doing something. Then you can do it. When he took his wife back in time he imagined creating a beam of light and attaching her to it and then letting her go. She went back until she found an obstruction, and that obstruction was that law. At that point the information was released as to what had happened.

He said, "Now, people call it cell memory. All the experiences from day one are locked in those cells. It is not the information that's lost; it's the access to that information that has been lost in a complex maze of doubts, fears, social blocks, mental blocks, and disbelief. Whatever can create that will stop that access to that information."

SEEING INTO MYTHIC TIME

As far as all of the shamans were concerned, it was only our belief and our judgment that kept us from free access to the past, even to past lives. To me this meant that time traveling required one to shift one's perception, one's way of seeing the present. I had felt such shifts, momentary as they were, when I first arrived in Lima. I had noticed the presence of history in the faces of the people I saw in the streets. This was a clue.

Normally we become accustomed to our surroundings, our friends and lovers, our everyday life. That is a hindrance to time traveling. As Kote had told me earlier, we either have completely lost five of our ten senses already, or are close to losing them. These five are just as important as the five we still possess. You'll recall that these are the senses of imagination. In a new surrounding, meeting new people, or falling in love, these senses hopefully are awakened. When they are we enter into a timeless state.

I thought of mythic time running perpendicular to chronological time. Thus, in a real sense, mythic time is timeless. There are experiences to be had, visions to be seen, but they all take place in a flash. It was in mythic time that I believed one could have visions of the past and future.

I then remembered my conversation with Dr. Jacques Mabit a few weeks ago. It was Sunday afternoon just after my first ayahuasca experience in Peru. After I had awakened from a nap, Jose Campos, the young shaman I had met two days earlier, had arrived as he promised and with him was Dr. Jacques Michel Mabit.

Dr. Mabit, a young medical doctor and Frenchman, was living in Tarapoto, Peru, and had spent the past three years carrying out experiments with the *curanderos* of the Peruvian upper Amazon region. As I was speaking with him and Jose, I thought of my own paper[3] and about my theory of time described in it.

[3]Wolf, Fred Alan. "On the Quantum Physical Theory of Subjective Antedating." *Journal of Theoretical Biology*. Vol. 136, pp. 13–19, 1989.

According to the transactional interpretation of quantum physics,[4] it is possible that a future event could resonate with a present event in such a way that a flash of the future could be perceived in the present. To the perceiver of this message, nothing need appear out of the ordinary. He simply thinks that he had a new thought or, in fact, a thought perhaps of something familiar. Of course, under special circumstances, this could appear as something quite amazing.

In fact, our normal perception of the world, according to what I wrote in this paper, requires that we must receive such messages, just to know how to deal with the everyday world. However, which of the many possible futures do we receive in the moment? There are many possibilities, as you all know. How does one know which will be the real future?

The answer, of course, depends on how far we look into the future. The future just a billionth of a second from now is about 99.99 percent certain. We know what will happen with some degree of certainty if we don't project too far ahead. This also depends on what events we are looking at.

In this transactional interpretation of quantum mechanics, messages from a short time ahead are more or less certain to be received. Messages from a more distant future are seldom heard. The reason has nothing to do with the strength of the message but simply has to do with these other realities. The farther into the future one goes, the greater the differences in these realities. Thus there are many different realities sending back messages to this one. Consequently the message from the distant future is garbled.

A similar situation exists for the past. We can consider that there are many possible pasts. It becomes blurry if you go back too far. Indeed it even becomes blurry if you go back just a few seconds. If you ask people to say what happened to them, even if they were all sitting together and witnessing the same event, they will tend to describe some aspects of the event in common, but they also will describe many aspects differently, as if they were seeing parallel events that had some commonality but yet still had distinct differences.

SUBJECTIVE ANTEDATING

As I mentioned, the shorter the time interval over which you used your mind to time-travel to the future, the closer the information appeared to you as normal reality. There was some evidence that this was the case.[5] Benjamin Libet and his associates at the University of California, San

[4]See the Cramer references in the bibliography.

[5]See the Libet and the Eccles references in the bibliography.

Francisco Medical School, actually gained access to the cortical areas of people's brains during standard brain operations, such as the removal of tumors. He noticed, by using refined electrical mapping techniques, just when certain areas of the cortex, those associated with body areas that received stimuli, would fire. He called this firing pattern "neuronal adequacy." He showed that a cortical area associated with a stimulus would not show neuronal adequacy until a full half second after the stimulus had been applied.

For example, if he touched the thumb with a pin, the area of the cortex associated with that thumb would not fire until a half second had passed. And yet when he questioned the subjects about their experiences, they invariably believed that they knew about the stimulus—what was being done to them and where—instantly. In other words, at the same time that the stimulus was applied.

The question was, how could a subject be aware before his brain is aware? Further tests indicated that the subject certainly "knew" that a stimulus was applied in just a few thousandths of a second after the stimulus. The subject would push a button as soon as he sensed the stimulus. It is known that there is a neural pathway through the spinal column that short-circuits the brain so that a "knee-jerk" reaction can occur without the brain being involved.

But the question here dealt with awareness. The subject said he knew what was happening to him before his brain had time to process the data. Presumably the subject was "mistaken." He was simply antedating his experience, claiming that it happened before it actually occurred.

At least that was one explanation. Then our brains, supposedly the seats of our consciousnesses, are just passengers going along for the ride. They see the sights, feel the bumps, but they have no control. They think they are in control. They think that every time they see something it is because they somehow directed their minds to pay attention to the visual event. But if they are antedating, that focus of attention was an afterthought. They were being led by their noses to pay attention. But they believe that they had control and free will. If the antedating hypothesis is correct, then like free lunches, there is no free will.

In my paper[6] I offered a new idea concerning this subjective antedating. I suggested that the laws of quantum mechanics apply to the brain, particularly to the operations of vesicle emission. In this manner, there must exist probability fields in the brain. Now these fields operate according to quantum rules. This means that whenever a state is cognized in the brain

[6]See Wolf, Fred Alan, *Journal of Theoretical Biology* paper in the bibliography.

(something becomes known), these probability fields must undergo a sudden change. What was probable suddenly becomes actual.

But it is very difficult to explain just how this passage from probable to actual takes place. Using the transactional interpretation, this process becomes clear. Not only that, but the reason for the antedating also becomes clarified. Accordingly, the stimulus sends a quantum message to the cortex, which receives the message a half second later and then sends a message back in time to the stimulus.

In a nutshell, the future communicates with the present along the neural pathways in the body. Our body-minds are, in a sense, time machines. This would explain any precognition experiences, visions of the future, and would also explain why they are so rare. Those pathways associated with survival are the most probable pathways for messages from the future. The future brain sends back in time that message that guarantees that the body will continue to that future. It is a self-consistent time loop.

Jacques found this interesting. He mentioned that he had read a book by Julian Jaynes, entitled *The Origin of Consciousness in the Breakdown of the Bicameral Mind*. Although I hadn't told him, I had referenced this book in my own book *Taking the Quantum Leap*, many years before. I thought that the ability to have a will was a later evolutionary step that arose through a quantum mechanism taking place in the brain.

This breakdown of the bicameral brain was a quantum event—a step leading to a new form of consciousness. It might have occurred for Moses when he came down from Mt. Sinai. It might have been the break that led to individual consciousness. It may have been the first time the humans no longer heard the voice of gods, but only had their own to listen to.

DESCRIPTION OF AN AYAHUASQUERO

But now I wanted to know what ayahuasca could do to alter my consciousness so that I could time-travel under its influence. I then asked Jacques to tell me about how he prepared himself for time traveling with ayahuasca.

He explained that you must follow precise rules and prepare a special diet. (I described the importance of diet in Chapter 4 and especially in Chapter 5.)

I asked Jacques and Jose, who was very absorbed in this discussion, "What must I do to prepare myself so that I will be able to see into the future using ayahuasca?"

"The message from the future comes. You can't provoke it," Jacques said.

Jose responded, "It is totally spontaneous, you can't control it. It just happens and you see. But the diet is very important if you take ayahuasca. The song, the icaro, that the shaman sings is also very important."

"But I don't understand the songs," I said.

Jacques responded, "It is not important to understand the icaro."

I was concerned about the diet. I wasn't prepared when I took ayahuasca the previous Friday evening. I knew that it wasn't working for me as I may have expected.

Jacques explained that, when he says diet, he means that you must go to the jungle and stay for at least three, four, five days, or even one week. You remain on a special diet. You may take in no food at all. You should try it for three days the first time. He had done this, taking in only water. But generally, the shamans don't do this. They eat only green bananas and some kind of fish without salt. It is also important to refrain from sexual relations, to not use any perfumes, colognes, aftershave, deodorants, etc. You also must be very isolated from anyone else. It is dangerous.

This kind of preparation will be very conducive for bringing on visions of the future. The more you diet, that is, go without food, the greater the experience, the farther you go. This is the main point with ayahuasca. You can take it a thousand times, but if you don't take care with your diet, you gain no new insights. With some dieting you will have an experience. But with less food, you will have a greater experience.

I suspected that Jacques and Jose had indeed done this. I was hoping that they had gone far enough to experience some of the paradoxes of alternate realities, even parallel universes.

But Jacques said he now believes, after his experiences, that there are not a lot of realities, there is only one. The many realities we experience are really just reflections of the one reality. When you diet, you see this. But to see this, he had to work on himself, his problems, his psychology, etc. He experienced many obstacles keeping him from seeing this single reality.

What was this single reality? I asked him, "What did you see?"

He experienced what he called the only reality. It was a simple sense of presence, without future and without past. No time.

I felt that this was very confusing to understand. I knew that our minds were always bombarded with thoughts of the future and the past. I suspected that he was referring to the single presence that perhaps the Buddha had talked about.

Jorge, too, felt that there was no separation between times. There is just a preeminence or predominance of today over tomorrow and over yesterday. We fail to see these other time-worlds because conscious life is nothing more than a social agreement.

I saw that, by analogy, the experience of time that we feel is similar to how, at this moment, we are experiencing a time of day. We have a social agreement that keeps us locked into a single time experience, because this is how we want it and we all agreed upon it. A few time zones to the west of us, it is a different time of day. We move by consensus. In a similar manner, but still eluding me, it was possible to experience other time-worlds as we experience other time zones. They are there, just as it is nine hours later in Europe than on the west coast of the United States.

Jacques told me that he not only had visions, he had the other senses as well. He heard things and smelled things. He had all kinds of feelings. There was a crossing of his sensations.[7] He "heard" colors. He "smelled" sounds. For example, he saw colors of the icaro, the song.

He told me that there are two different levels. The first level was his life as a social person, his name, his country, his job, and so on. Ayahuasca helped him solve problems associated with this level—for example, problems with family, with the teaching of Freud. The second level is symbolic. You live within a world of symbolic references. You are not only a social person, you refer to yourself as a being. Your father is not only your biological father, he is also the symbolic father of all beings.

THE TWO LEVELS OF TIME

All of Jacques's work was now into the exploration of mythic time, as opposed to chronological time. He believes that this is the real time. Sometimes he realizes that he is living through both times at the same moment. He had become very conscious of this.

I asked him about specific visions when he time-traveled.

Sometimes he saw animals, maybe persons. For example, during one session he saw two French persons. He knew their names but he knew that he didn't know them. He didn't understand what it means when that happens. Sometimes he would see sick people and he would be curing them. He didn't know if these people really existed or not. He said, "I don't know if I will really cure them. I can see people that I know from television, public figures, and I see how to solve their problems. If I can solve my problems, I can solve their problems. The other and myself are one. There is no separation. Their problems are exactly the same as my problems. There is no separation between me and the others.

"In this sense, there is no other one. I am the other. I keep my specificity. But I am not the other. It is difficult to understand."

[7] I described this crossover phenomenon earlier in Chapter 5.

I knew what he meant. In the chronological time, you are separate from everyone else. In the mythic time, you are everyone else, there is no other.

He said, "During the sessions, I experimented with all of my emotions, all of my soul. I saw this. I saw the resistance that I had. The vision is everything. There is some kind of proof, but it is difficult to prove this kind of understanding. There is a lot of suffering when you take ayahuasca."

I knew that this was true. I wasn't the only one. He suffered, too, when he took the substance.

He said, "Yes, there is even more suffering the farther you go. The deeper you seek, the more you suffer but the greater is your gain, your insight, your vision."

VISIONS, TIME TRAVEL, AND THE UNCONSCIOUS

Shamans were visionary. And now I suspected I knew what these visions were and where they came from. I knew, of course, that, as a rule, we Westerners seem to not be visionary. But why didn't we all see the way shamans see? Perhaps we did, but the messages are encoded because they involve a superposition of many possible waves from the future overlapping with images from the past. In other words, we would see them as possible futures and pasts all overlapping, forming a very blurry image at best. I believed that shamans knew how to reconstruct those images in terms of archetypes.

But how would these archetypes arise? I now had an idea. Perhaps these archetypes were stable patterns within the unconscious mind, much like the energy levels of atoms are stable patterns of energy. It is really amazing that after two atoms have collided, they revert to the same indestructible, stable, orderly, energy patterns. If two automobiles collided, nothing at all like that would happen. After an automobile collision, a disorderly mess remains. But if we could see that collision in terms of atoms, a very orderly pattern would remain.

Indeed after an airline disaster, investigators put together bits and pieces to reconstruct the accident. They, in a sense, see what actually occurred at the time of the crash because they know how to look for the memories of the disaster.

At the ultimate level of reality, atoms, there is also a form of memory. It is contained in their energy patterns. Perhaps our unconscious minds catalog images of the future and the past by breaking them down into similar energy forms that we call archetypes. I suspected that these archetypes constituted the orderly memory banks of the shamanic world.

But we normally fail to see these archetypes. We do not have the decoding device that unlocks the bank of archetypal memories. Just as it took science many years to decode the images from the atomic world, it will take us some time to distinguish images from the shamanic world. We would need to learn how to observe these images. Much as a person undergoing psychological training learns to see certain images, depending on his particular state of consciousness, in a Rorschach test, the observer of archetypes would learn to see his own "screen of consciousness." This ability could be explained as part of the observer effect in quantum physics, only this time the observer would be able to resolve images based on perceptions of superpositions of the past and the future in terms of stable archetypes.

We all have memories of the past and fantasies about the future. But which images did we choose and why did we choose them? In other words, what laws of the universe are in effect that cause us to choose one set of images over another? There had to be certain images that fit into what Jung called the "collective unconscious." Otherwise, these images wouldn't be so persistent in a culture.

Somewhere along the way of our history, we seemed to lose that ability to see ourselves in terms of these archetypes. If we could awaken that ability in ourselves in this everyday world, if we could remember ourselves in terms of archetypes, I felt that it would open our hearts and minds to a new level of shamanic consciousness that could lead to a new level of health and well-being.

Freud had pointed out that in the unconscious mind, time order made no sense. Jung believed that the archetypes of mythic reality lay in the collective unconscious. Each of us had the memories of all of us, only they were categorized in terms of these archetypes. Joseph Conrad once said, "The mind of man is capable of anything because everything is in it. All the past, as well as all the future."

Candace had pointed out that our consciousness is divided into two parts. As I saw these parts, there were an overpart capable of willing change and an underpart incapable of being accessed by the direct will. The underpart was the body-mind. It had been well schooled and taught just what was necessary for survival. It included unconscious desires that the overpart, even though it wished to change them, was not capable of directly changing. It was in a sense "addicted" to its behavior patterns. It believed that the patterns of its own desires were necessary for survival, in spite of the evidence that may be contrary to that.

We certainly needed that underpart. Freud called it the "unconscious." But it really is not unconscious. It is very conscious, but it is locked in a rhythm, a pattern, that it believes is necessary for its survival. Of course, it is correct most of the time. It tells us when we are out of breath or in need of food or water. It tells us what to do in moments of stress or impending danger. It says to us, "Be aware," something is out of kilter.

But it also is capable of intoxication. It can become addicted to substances and past experiences. If it is convinced, by whatever experiences it has acquired in a single lifetime, or in many lifetimes, that a certain way of behaving is survival-enhancing, then it will send messages of desire to the overpart of its needs.

The boundary between the overpart and the underpart was a blurry surface. Freud called it the preconscious mind. Things from the underpart would bubble into the preconscious boundary and be picked up by the overpart as desires.

This boundary region interested me. I thought that the shaman has learned to reach into this boundary region to access the memories of the body-mind, the underpart. But to do this the shamans had to develop a technique. How did they do it? What was it that enabled them to enter into this region so well, and with so much of the overpart present?

The answer was there. But it frightened me to think about it. It was an ability to transcend the death of the body-mind. To somehow, by the intent of the shaman alone, to convince the underpart to be an ally, to be an accomplice, in its own death.

IO

Death

Meliton explains to Miguel that, in the jungle, Nexy had to confront the death of her mother, and the shock that this created when she was a child. He tells him that the death of Nexy's mother created her inability to function in the world. He asks Miguel to look into his past and to see what is haunting him. Miguel realizes that he has been afraid of death, refusing to acknowledge it. He remembers the death of his own father. In his father's death he sees his own passing. He begins to see that his own death is not just a final process, but one that continues through his life. He sees that he has been avoiding death, not realizing that his own life was to come to an end. And with that simple realization he realizes that his life must have a deeper purpose.

As the scenes of Miguel's realization appeared on the screen, I began to recall my own experiences with death and the connection that this experience had with shamanism. My ninth hypothesis was that shamans enter the death world to alter their perceptions in this world. I had thought about death many times. Even my physics training dealt with death.

Although it sometimes seems to be a fantasy, the new physics of today's world is a far cry from the old physics that most of us were taught in school. In the old physics, there was just one universe to live in, and this was it. It consisted of countless billions of tiny dead solid bits of matter interacting and exchanging energies with each other, and somehow out of this, life arose. But the prevailing belief behind this was that the whole universe was essentially dead and just going through the motions.

The new physics had, behind the present technological education of most Westerners, changed this somewhat. No longer was there the vision of just one universe. As I mentioned earlier, modern physicists use the concept of parallel universes, an infinite number of them, in their calculations today. Not only that, but for some physicists, matter itself is no longer seen as dead. It is, itself, not even solid. The tiniest bits of matter are more like ghosts or winds of possibility that somehow manifest as grains or bits of stuff whenever they are observed.

My mind shifted between the two schools of thought. In the old view, time and space were infinite and unimaginable. There was always more space to conquer and enough time to do it in. The universe was seen as bountiful in its infinite amounts of dead matter. In the new view, space, time, and matter are inextricably bound together. Time may be traveled through both backward and forward, and space transported across taking any amount of time to do it, and life is present everywhere.

Why is death important to the shamanic experience? There were several reasons.

Although it might appear a little droll, death was the only way that one could convince one's body-mind that an altered state of consciousness was necessary. This had been pointed out to me, when I had visited England last year, by the Druid pendragon, Jerome Whitney. Jerome had told me that when the body is dying or is convinced that it is, the body-mind shifts its mode of perception.

Shamans do not need to die to do this. Chris Hall and Richard Dufton, the shamans from Brighton I had met last year, had also told me that they recognized the shaman's use of rhythmic instruments to achieve vision of the death world. Somehow when a drum beats or a horn blows as in a Buddhist temple, death is at the door. I don't know how this happens. I feel that it must happen, if an altered or shamanic state of consciousness is to be achieved. Perhaps this is what the vibration was needed for.

When the body is convinced that it is dying, the mode of reality perception shifts. It must, for the usual mode of perception, the one that we have all acquired in life, is no longer carrying out its prime function—keeping the body alive.

In my model of reality, our sensory apparatus perceives more than we actually are aware of. What is perceived as nonthreatening to us is simply ignored. It is there, nevertheless.

In our society, there are so many obvious "real" dangers. We live in a highly artificial situation. We have dangerous means of moving ourselves at high speeds from one place to another. In an airplane, even though you manage to fall asleep or take a nap, you are always aware of the danger possibility. The same is true in a car moving along the highway, or even in a house with electrical appliances around you. That danger is mollified quite a bit. Each moment that passes without something bad happening to you soothes you into a false sense of security.

But the danger is quite apparent. In the jungle or on the farm away from technology, those technological dangers vanish. It takes about a week for them to leave your consciousness, your body-mind, if they leave you at all. Then you begin to sense new possibilities, new dangers. These can be the sense of other animals around you, or the possibility of falling and hurting yourself when alone in the jungle, or even the possibility of being lost if you wander too far from the village.

In other words, wherever you happen to be, your body-mind will orient itself to keeping you alive. It will attempt to sharpen those senses required to maintain your health and well-being. If you are ill, disturbed or preoccupied, or lulled into a false feeling of security, you can, of course, interfere with this natural protection. Most so-called accidents occur when we are so preoccupied. The unexpected takes place. I walk by a tree and a branch scratches me. I move my car onto a crossroad, and a car hits mine. I stop to make a left turn, and the car in back of me rear-ends me.

So when death threatens, the mind begins to search for new ways to survive and pays attention to things that it had ignored. Thus when a shaman visits alternate worlds, he is threatening his life and forcing his consciousness to deal with the unseen world of his or her own death. The mind seeks what could destroy it.

And now my mind was seeking to understand the connection of death to life itself. And I was remembering how close I may have come to my own death. Suddenly, I felt frightened. Although I have only told you about four so far, I had been through five ayahuasca sessions with shamans. When I had taken ayahuasca I had become very ill. I hadn't realized it then, but I had been in some danger without knowing it. Perhaps it was better that I didn't know it at the time.

How could I have been in danger without knowing it? Somehow I had become used to these ayahuasca ceremonies. But how did my mind enter into that state of acceptance? Was I accepting my own death? Had my mind fallen into a rut, so to speak, of false security? Or was this an essential clue to living a fuller life by accepting one's death?

I recalled the events just prior to my fifth ayahuasca session. They took place just a few weeks ago when I was in the jungle. I was waiting for Don Solon's return from Iquitos and had decided to take a walk with my guide, Franco, along the shore of the Amazon.

. . .

Franco and I looked for a canoe, but there was none to be found. We had hiked about a half-mile from the hut, moving in the direction of Iquitos, upriver. If we had found a canoe, we would have then canoed downriver, returning in the direction of the retreat. We stopped along the way meeting various folks who lived along the river in open-air surroundings, very primitively.

I was hoping that we would find a canoe, and if not, at least a boat. I had become exhausted from the hike. The heat was ever increasing, and each step was an ordeal. You don't just walk in the jungle, you climb. You lift each foot over that vine or this tree trunk, or around that mud hole. Or you stop for a while and cut away some growth to make your way. A half-mile's hike in the jungle was as exhausting to me as if I had climbed a half-mile up a mountain. It felt the same to my body.

Well, we didn't find a single floating craft. So Franco, seeming not to mind the hike, said we must walk back. We did and I hated every minute of it. I was sweat-drenched when we finally returned to the hut. I stripped off my clothes and went into the Amazon to cool off.

My ordeal in the jungle made me remember all of the ordeals I had been through in the past few years. I remembered the firewalk and how all of my family had undergone the experience. However, my ordeal with fire hadn't ended with the firewalk. My mind drifted back.

By the winter of 1986, I had felt what I am sure all my fellow writers feel from time to time, that my writing career had come to an end. I began to have second thoughts about my remaining years and whether I had made a good decision to leave the academic world. I had noticed that a teaching position in the physics department was available at the University of Northern Iowa. I contacted the chairman, and was relieved when I was offered the post. It was a temporary position and I was eager to see if I still had that old zest for teaching physics in the classroom. Judith was apprehensive and felt that this wasn't what I was destined to do, but she was willing to go along with it.

We left La Jolla in January. My son, Michael, came to see us before we left. I have always had a strained relationship with my children. Divorce and separation from them has never made our being together very easy. But Michael was different. He made it a point to keep contact with me, in spite of the rift that split me from my first family. Consequently, or for no reason at all, I loved Michael very deeply.

Michael was a difficult child to raise. He always had a deep interest in what I would call the dark side of nature. From the time he was nine years old, he managed to wound himself, losing his left eye in a BB gun accident. In his teenage years, he broke an arm.

Michael was also a drug addict. This was the hardest blow for all of us to take. He had been involved in street drugs ever since he was nine or ten years old. Living in the lotus-land of La Jolla, California, he had nevertheless found his way into the dark underworld of the criminal and the drug addict. The two are never very far from each other.

When I had gone to Europe in 1974, Michael came to spend five months with me, attending school in London. Once I had taken him to the west coast of Ireland and the Aran Islands, a place I love to visit. The isolation and the great beauty of those islands had a special spell. Michael and I had walked to the western edge of the center island, a few miles from where we had taken rooms at an inn. When we reached the edge, I noticed that there was a sheer drop of several hundred feet to the sea below.

The waves were violently kicking up, spewing foam over the rocks. Michael and I looked down. I was frightened and backed off. Michael was delighted and moved closer. He was only thirteen at the time, but he showed no fear of falling off. I cautioned him to move back. But, as always, danger held its fascination. Michael picked up a huge flat rock, weighing at least twenty pounds, and hurled it from the cliff down into the crashing sea below. The rock split and exploded as it hit the roaring ocean and the crags jutting up at the base of the tor.

I suddenly felt a chill. I'll never forget it. I saw then that Michael was fascinated with death, a fascination that he never lost, and I felt that Michael wouldn't be with me much longer.

Michael's experiences with the drug realm were adventures in death. He became a heroin addict when he was no more than sixteen or seventeen. He had probably taken the substance before that time.

Several years later when Judith, Shawn, and I had taken our first home together in La Jolla, Elaine, Michael's mother, had called me and told me that she couldn't handle Michael's drug problem any longer. She asked me to take it on and have Michael move in with us for a while. He did.

Have you ever had to deal with the problem of drug addiction? It was an amazing lesson for me. Michael promised that he would stay off drugs and remain clean and sober. I believed him. Judith did not. I was too naive about the power of addiction. When you are addicted to something that does not have the approval of those who love you, you lie, cheat, and steal. You do anything to keep the family off your back. Judith wasn't fooled, and she knew what I was not able to even sense.

One evening Michael said he wanted to go for a walk along the ocean near where we lived. I said, sure, why not. As soon as he left, Judith said, follow him, he is going to get heroin. I couldn't believe it, but we decided to follow, like detectives, and observe what he did. He went to a telephone booth and was talking to his drug contact as we walked up and completely surprised him.

We took him home and, after much screaming, crying, and yelling, got to some of the truth.

As time went on, Michael began to see that the only way he would escape the addiction was through his own sacrifice and belief in a power greater than himself. He managed to stay clean and sober for three years. He worked in a drug rehabilitation hospital. He lectured in many places about the different drugs and their effects. His knowledge of drugs was enviable. Often doctors running drug "rehab" clinics would ask him to speak before their groups.

Michael had found his metier. But the fascination with drugs was hard to shake. Michael once again got hooked. But this time his suffering was profound. After six months of rehabilitation, he seemed reconciled to spend the rest of his life off drugs. When we left La Jolla for Iowa in January, Michael was in good spirits, and I was very sad to say good-bye to him. It was the last time I was to see him alive.

We arrived at Cedar Falls, Iowa, in the winter, and Judith and I were a little disheartened to be there. I was born in Chicago and knew what midwest winters were like, but for Judith it was a profoundly shattering experience. Nevertheless, the faculty of the university welcomed us warmly, and soon I had settled into the familiar routine of my teaching duties.

In February we decided to take a weekend off and visit with my sister in Chicago, 250 miles east of us. That Saturday evening, after having a meal at one of the many good Chicago restaurants, we went to bed. Suddenly I heard a strange noise outside. It seems that the young man living with his parents next door to my sister had a severe alcohol problem, and he had just come home drunk. He had parked his car, but in his inebriation, he managed to crash his car into the car parked in front of his. His father and mother had come out to the street and were yelling at him. I didn't think much about it, although I had taken note of the time.

The next day my sister, whose name is also Judith, my wife, and I decided to brave our way into a violent winter storm and go to see a film. It was the movie *El Amor Brujo* (Love, the Sorcerer), a brilliant film by Carlos Saura. The film dealt with love and death in a small poor community in Spain.

When we got home, my niece Liz told us that my daughter Jacqueline had telephoned with awful news. I immediately called her and found out that Michael had been struck down and killed Saturday night by a drunken driver. He was simply walking along the roadside to a restaurant with a friend who was also a recovered drug addict. Both he and his friend were killed. It was terribly ironic that the night he was struck, he and his friend, after their dinner, were going to an Alcoholics Anonymous meeting to help others. He had been working with other addicts, including alcoholics, helping them up to the moment of his death. He had been clean and sober for just about two years until that tragic night.

ABSALOM

Nothing can describe the horror and pain of the loss of one's child. It was the most shattering experience I had ever gone through. Suddenly a world stops, and my mind filled with all of my experiences with Michael. And then the sudden realization that the very night, perhaps at the very time that Michael's world came to an end, the young man next door had crashed his car below our bedroom, stoned out of his controlling mind on alcohol, was a bitter synchronicity. The eerie film just added to the mystery of Michael's sudden death.

We flew the next day back to La Jolla and joined my estranged family. We never felt closer than in those days of mourning. Michael's death was not splashed over the front pages of newspapers, although the announcement of his death made the local television news.

We decided to have a funeral at a local church in La Jolla. I had no idea of what to expect. We were all shocked. The funeral was attended by no less than five hundred people. Perhaps an expected number for a famous person, but hardly for a boy in his mid-twenties. I was amazed and deeply moved that so many came and that he had touched so many people's lives in ways that were truly revealing to me.

The ceremony lasted over two hours. Perhaps one hundred people got up and said something about their relationship with a son I had really never known. Most of them were addicts. Yet what Michael brought to them was a sense of love that none of them had felt before. One young man said, "When I was suffering, Michael turned to me and said, 'Look, I love you.' Then Michael embraced me. I was shocked because no other man, including my father, had ever embraced me and told me that he loved me. This opened my heart. Because of Michael, I was able to go on and help others."

Michael had taught many about love. He had a gift that was of his name, angelic.

THE VISION OF MICHAEL AND THE HOLOGRAPHIC UNIVERSE

Michael was twenty-five years old when he died. About two days after the funeral and after Michael's ashes were buried at sea, a wish he had told his friends, Michael came to me in a striking lucid dream. We talked. He appeared to be happy and relieved. I was certainly relieved and joyous to see him. He seemed so real to me, but I knew I was having a conscious lucid dream. So I reached out to touch him. I grabbed him by the ankle to make sure that he felt solid to me. And, as in all my lucid dreams, since my sense of touch and solidness remains intact, he felt real. I could feel the thickness of his ankle and even the hair on his lower leg. I wanted to reassure myself that somehow Michael was still living. Michael then communicated to me, without words, "Dad, all your theories about the holographic way in which the universe is constructed are correct. You are seeing it correctly. It is a gigantic hologram woven in space-time. It is no more real than this or anything else."

I had the impression that he was as real after passing from this world, in his world, as I was in this world. I realized that neither world was more real than the other. We were both part of the warp and weave of the giant hologram. Michael gave me the impression that if I needed him, he would help me. He would be "there" if I needed him.

Over the past few years, I have felt Michael's presence. Not in any spooky or weird way. But if I am presenting a talk before many people or I am undergoing a sacrificial act such as in a sweat lodge, I feel he is with me.

I later had other dreams of this special person who had been in my life in the guise of a son. In the last dream I had of him, a few years ago, he had undergone a transformation. He appeared indifferent to our world, even though he recognized me in the dream. I was riding in a car with him. He was sitting alone in the back seat and I was in the front. I had turned around to look at him, and his whole body was glowing. Michael was undergoing transformation for the next part of his journey into the great mystery.

Michael had made me aware of death in a way I hadn't expected. I became fascinated with the great door separating lifetime from lifetime. My teaching experience at Northern Iowa had little impact on me. I was renewed in my belief that I had to go on with my life as a writer, speaker, and teacher. I had to be fearless in the face of adversaries. By sticking my neck out and writing about physics as I had done, I had already raised the scorn of many critics, especially some fellow physicists.

I didn't care. I sensed a new importance to the information I had to share. I was no longer interested in just teaching and explaining physics to others. The universe was far richer than a few simple myths of science. Even quantum physics was too limiting. There was another world after death. I was convinced of it, in spite of many psychological-Freudian-Jungian theories about the state of consciousness of a mourner or the grossly simple theories that dream content is a waste of time and just the images from random electrical discharges in the brain. I had studied the deepest issues of science, the King of the Sciences, Physics, and found that the emperor was shockingly devoid of any clothing, even though many are in awe of his regal attire.

THE PENDRAGON AND THE MOVIE REEL OF DEATH

I knew that all shamans dealt with the death world. It was as real to them, perhaps, as this world was. I was hoping to solve a problem that has haunted humankind ever since the beginning of time and the first recognition that no one of us ever has but a finite piece of time to enjoy or suffer through. I wanted to know if death was the final answer, the final end to conscious life. It seemed to me, from my past experiences dealing with the deaths of those closest to me, and from the dreams I had after their passing, that death was not the end by far. If it wasn't, then what was the end, and as all of us are inclined to ask as our numbered days pass, why are we here? What is this experience for?

I found some answers to my questions from the American pendragon of the Druid Order, Jerome Whitney.

I had told him about my dreams, especially the visit to the astral plane of souls that had aborted their bodies through suicide. Jerome told me that there were esoteric students in the 1930s, and even later, engaged in what they called rescue circles. As far as he knew, there still are rescue circles in England. A rescue circle meets as a ritual to establish contact with people who have passed on. They want to help people whose names are put before them—people who have been caught in a rut, having died, and need to be nudged on. Rescuers want to provide instruction for people to move on.

The problem with our dealing with death in our Western culture is that there isn't an established procedure in general for the masses for helping a soul after it has died. In the East, the *Tibetan Book of the Dead* gives instruction on how to provide for the spirit after it has left the body and is in a disorientated state. When a person dies, he has to be very evolved and powerfully orientated to be able to move out and not be disorientated.

Jerome explained that the *Egyptian Book of the Dead* (different from the Tibetan) also contains instruction for the living. These instructions tell the living what to do to aid the dead person on his journey. Catholics have a similar procedure for the Pope. He actually receives instruction after he dies. People around him have a little rubber hammer and they tap him on the head three times and they talk to him. This procedure, using the little hammer, is a shamanic carryover for instructing the spirit. But in general, when Joe Smith dies, there is no one who knows what to do, besides call the men with the zippered plastic bag.

Jerome's thoughts were sobering. I realized that in the United States, there is a deep-seated anxiety that no one survives death, in spite of all the religions we have. Our ceremonies are really for the living and not for the dead person. What I was realizing was that we need to treat death as a doorway. The person who dies is, in a real sense, not dead. He or she still exists, but not in this part of the universal hologram anymore.

There is another aspect of a hologram, as a metaphor for the universe, that I need to explain. You'll recall that a hologram is made by shining coherent light on an object and at the same time through a film emulsion. The light that reflects off the object plus the original light both strike the film emulsion, leaving a wave interference pattern. That pattern is recorded. When light again illuminates the hologram, the whole object appears in a completely three-dimensional form. In fact, as the viewer moves his head and viewing point across the hologram, the object appears to the changing view much as a real object does. For example, if you move closer to the hologram, the object will tend to get larger, just as a real object appears to get larger when we move closer to it.

Now, there is another exciting aspect to a hologram. It contains multiple images. By changing the light source that illuminates the hologram, different images will appear. The hologram has encoded in it a series of possible images. What one sees depends on how one looks. You can think of the hologram as a composite of parallel worlds, each seen only when illuminated properly.

The spirit, like the hologram itself, exists everywhere. It appears in one form or another depending on how it is seen. However, like a hologram you must know what to look for when you experience a spirit. More to the point, you need to have developed senses that extend beyond the physical. In a very real sense, we, in the Western world, have lost our senses that extend beyond the physical. Since we have lost them, we deny that they even exist. But even worse, we deny what those senses could tell us, if we still had them.

Jerome pointed out that this is quite evident in the ceremonies for the dead. When people in the United States die, there is, of course, a lot of weeping and mourning. The person is buried or cremated. But what effect does all of that have on the person who has died? Perhaps that grief anchors the dead person and holds him spiritually from evolving for a needlessly long time. Jerome senses this when he goes into cemeteries in the United Kingdom, where there are recent burials. He still feels the family's grief holding this poor spirit from moving on and unfolding the experiences of the past life and building them into the next.

Jerome told me that Thomas Maughan, the past Druid chief, had taught that in any of the near-death experiences, people start seeing things backward. They start seeing their life unfold as if they were watching a movie. The movie reel of your life unwinds when you die.

While I knew that all shamans pass through near-death experiences before they become shamans, I wondered about the pendragon's experiences with death. I remembered our meeting once again.

THE MOMENT OF DEATH

Jerome watched my face for a moment and then told me that he had a near-death experience in 1975. He had smoked about seventy cigarettes a day for many years and was running down a hill too hard. He overstrained his heart. As a consequence, any time he overstrains his body, which he was starting to do foolishly nearly every year, his body-mind would get afraid that it was going to die. He said that he doesn't worry too much about it now, but still his body-mind worries that it is dying. Several times when he had overstrained himself far too much, there was a beginning of a sudden recapitulation of all of the experiences that were gone for many years coming back. They started flooding back in.

He said, "One particular time, I had no reference point for body sensations. My body went through a whole series of events that could not be related to any experiences of my past. There were feelings and sensations produced in it that made it go wild."

I asked, "What brought this on?"

He laughed when I asked him and said, "I was quite sick and wanted material to come out of all of my orifices, all at the same time. And the body could not relate to that. And so it decided that it was dead. The mental brain decided that I was dead. You see it hadn't had that experience before. It's a fascinating thing. The mental brain relates to the past. If it has no past experience and it can't see a future, from its standpoint, it's dead."

He continued laughing as he recalled the experience. He told me that it wasn't funny at the time. The body-mind was very afraid. But when that happens you experience this unwinding movie reel of your life. This experience is also described in the *Tibetan Book of the Dead*.

I wanted to know if there was any sense of time during this experience. I was beginning to theorize that near-death experience could teach us something about the nature of time. When one dies, one is in a sense freed from time barriers. If true, this might explain some of the ways that shamans accomplish their tasks. Perhaps they can move through time as easily as we move through space. Jerome told me that his experience didn't feel like time. I asked him, "Since it was like watching a movie, did you see yourself in the movie or were you simply watching from a distant viewpoint?"

He said, "It's not visual in that sense. It's awareness. You become aware of the event without emotion. Now this is interesting because there isn't any emotion attached to it."

He then returned to a lesson he had learned from the past chief Druid. He said, "In the Druid teaching, as taught by Thomas Maughan, the reel has to unwind seven times. In other words, you die, you then sit in what he calls the great movie show in the sky and you watch your life unfold from effect to cause, so you see it backward. With your increased awareness you first see the result and you then see the cause of it.

"This is how the learning of a lifetime is built into the spirit. Because the body has been a recorder, recording everything during its lifetime, recording everything on different levels. The mental, emotional, and spiritual levels have all been recorded. Thus, dying is like a debriefing. It's no different from the debriefings the astronauts had when they returned from space. You are just like the returning astronaut. However, the reel runs backward, each of the seven times at a particular level.

"But then it goes up each time to a higher, more subtle level. So you start back and you get an etheric unwind, and next an emotional unwind, and then you get a mental unwind."

Jerome's description was revealing. I was beginning to piece together the connection between quantum physics, time, and the death experience. Our sense of time is personal. Although we have, in physics, adopted many standards such as the second and the minute, and even though these standards are based on very objective observations such as the rotation of the planet Earth, the frequency of a microwave emission from an atom, or the orbit of our planet about the sun, our personal experience of time is quite variable. With your finger on a hot stove, one second seems like a long time, but in the pleasurable experience of making love, one minute seems like a few seconds. Our personal measure of time is based on the relationship between some internal clock, such as the movement of neural impulses, and the external world of events that we experience.

When your thoughts are rushing by, your internal clock is running quite fast. Thus after a mad rush of thoughts, only a few moments will have passed while, to you, it may have seemed like hours. When you are in a meditation state, the internal clock is running slow, so a few moments to you could be more than an hour of external time.

I believed that this subjective or personal relativity of time had nothing to do with Einstein's relativity. Physical time distortion, such as time-slowing, would only manifest when objects are moving at speed near to the speed of light.

But based on the data, when death approaches, the internal clock must speed up. So even though only a few seconds pass, the experience internally seems like years.

This much I understand. But what had me baffled was the fact that time ran backward during the near-death recall. The only clue I had was that the body-mind believed that it was dying. Could that have any effect on the way time was experienced? I knew that if an object could move faster than light, it would be possible for it to travel backward through time in relationship to the external world of objects moving slower than light. I couldn't see what would cause such an experience. As far as I knew, nothing moved faster than light. Nothing solid, that is.

Einsteinian relativity taught me that observers in relative motion would measure time and space intervals differently. For example, if two events, spaced one mile apart, were seen to be simultaneously occurring for one observer, a second observer, moving very near the speed of light, could see one of the events (event A) taking place years after the other event (event B), and the distance between the two events, as observed by the second observer, could be light-years apart.

If that moving observer happened to be moving in the opposite direction, everything would be in reverse. He would see the event B occur first and event A take place years later.

But this time reversal was possible only if the events in question were capable of being observed simultaneously by some observer. As far as we know, the events of our lives take place in time, one event following another. There was no way that anyone could see all of these events occurring simultaneously. In fact relativity was truly relative. Two events, such as the ticks marking off one second on a clock, would, for any moving observer, seem to be taking longer than one second. The moving observer sees the resting clock as moving by him just as you, sitting in your car before a stoplight, might see the car standing next to you moving forward and not realize that you were rolling backward.

No, the relativity theory didn't seem to be helping. But relativity has two parts to it. There is the dynamic part that tells us what is true in regard to the motion of objects. But there is also the purely kinematic part that only deals with observers in relative motion to each other. It doesn't say anything about the mass of an observer, only what an observer would see regarding the time and space intervals separating events.

So what if the observer *was* moving superluminally—faster than the speed of light? According to relativity theory he could not really have a rest mass. In other words, he could not be made of matter as we understand it. He would have to be an observer, capable of seeing, and yet not be made of anything physical. Could this be indicating that what was moving was not physical matter, but the movement of the soul itself? The idea fascinated me.

The equations of relativity aren't normally applied here, but they can be. A time interval of one second as seen by an observer going faster than light changes its form. It becomes a spatial interval. Now that is hard to imagine. And a positive time interval, seen by a resting observer, turns out to become a negative spatial interval, as seen by a superluminal observer. So a witness moving faster than light would see the normal flow of time as a sequence of events moving backward through space. If he were to move faster than light forward through space, he would witness events as if he were watching a movie running backward.

But witnessing these events is the wrong term to use. Because to witness something he would have to see it, and that means he would have to experience it as light. But he is going faster than light. That means that he really wouldn't "see" these events, he would have to sense them in some manner. Sight is probably not the correct sense. Could this involve the other five senses that the Chumash medicine people told me about?

I was coming up with more questions than I had answers for. Since there was nothing in physics that would allow any material object to move that fast, I was treading on thin ice.

If this was real, the soul would have to move out of the body faster than light. This opened a whole new doorway to the world of the shaman. Death was a movement of the observer, without matter, faster than light. What could this all mean? I was still missing something.

And, I wondered, was the movie I was watching a chance for me to see my own life in recall? I felt a strong connection with Miguel and Nexy, both. Each was reflecting to me my own life and, in a strange way, my awareness of my own death. After seeing her dead mother, Nexy was taken back in time to the scene of her early childhood. There she relived her mother's death when she was a child. And then Miguel saw the death of his own father when he was young.

As I watched I could only reflect on the connection between childhood and death. In some way children are closer to death than adults are. It dawned on me why this was true. It had to do with the connection between birth and death. I tried to remember what the Brighton shamans, Richard Dufton and Chris Hall, had told me about this. It was becoming a jigsaw puzzle. A piece here and then another seemingly unrelated piece there.

CONDUCTORS OF THE DEAD

We had been talking about birth and death when Richard added, "Remember that a mother is the best help to a woman giving birth, because she knows." Then I realized that the shaman acts like the antipode of the midwife. The midwife sees to the birth of a new soul. The shaman deals with the death of the soul. Shamans are conductors of the dead. They are the pathfinders through the maze of the death experience. If indeed our sense of time is blown completely by dying, it is no wonder that a soul would become confused upon dying.

To conduct someone means that you know the route. You have been to death and back. Shamans who were already technically dead are able to come back to life. But something occurred to restart them back again. Richard then said, "I 'died' of malaria. I had a 109-degree Fahrenheit temperature. I was gone for three days. White light. The whole lot. That didn't worry me. It was something tangible. But then something occurs in your life that takes your life outside your reality. I had recurring malaria. I used to go into coma. My spirit would go elsewhere. I learned things I had no right to know. And because I had been to the threshold of death seven times, I had become a psychopomp. Unless you understand death, how can you conquer it?"

Chris said, "And also, you can't conquer death. The most important thing about it is . . ."

I interrupted, "You see death as an ally." Chris agreed.

Richard then said, "When death doesn't tap you on your left shoulder, he's giving you the best gift of all, your life. The aspect of death can be 'female.' We say that the female is deadlier than the male. But when you're reborn, that action is patently female. Consider going down the white tunnel business exactly as a shaman does, like foxes down their holes. When I'm going down that, in point of fact, I'm going down a vagina. It's exactly like being born again to go into the new world. The most important thing for us to do is make a good death. Not necessarily a glorious one but an easy death. It makes it easier for your new mother, who has less pain to get over with, and then your growth period is sooner activated because most of your life is spent getting over your birth trauma. So if you make it easy on your mother, your life becomes easier, your growth is more active, and that is what the psychopomp is all about.

"I should have been dead years ago. I'm not, so every day is a gift."

DEATH AND BIRTH ARE ONE

Richard's description was another clue. Birth and death are truly one. You die and, in what seems like no time, you are reborn. Your experience in the Bardo land—the place that souls go to between death and rebirth—is timeless as far as the regular flow of time is concerned. Being without a body means that the soul has no mass. This frees it from all inertia and frees it from time.

If a person dies and then spends a timeless eternity in the Bardo land, moving in imaginal time, perpendicular to real time, he could be reborn at the very moment of his death. Thus the memory of passing down a tunnel could indeed be the simultaneous movement through the vagina into a rebirth.

The idea was fascinating me. I hadn't heard anyone say this before. As far as spiritual teachers were concerned, "heaven" was another place and rebirth always took place after death. But in this new vision, this need not be so. I was even wondering, if time was not a barrier to the soul, then was it possible for a soul to be reborn even before its original birth? If so, this meant that a soul born 100 years ago, for example, could have buried in its genes, somehow, memories of his future death in another life. In other words, he could be remembering a future—a time yet to occur.

Like Miguel in the movie, although I had witnessed the death of those whom I loved, I had never had a near-death experience. I wondered if the descriptions given to me by the pendragon of the Druid Order and the shamans of Brighton were close. I was particularly interested in the connection between death and time. I began to sense, in an uncanny way, that the physical time sense is what we experience as life itself and that, when we die, we are free of physical time and live in mythic time. This was the time of the shamans.

I remembered what Jorge had told me about his near-death experience when I first met him.

Jorge told me that when one of the patients of another shaman he had been studying with was severely ill, he was brought to the shaman's hut. The person was almost dead. He was in convulsions. He had a terrible amount of damage in his body. When the shaman and he took all the negative force out of the ill man, Jorge became unconscious. For forty minutes, more or less, he remained that way. He had no record of what had happened to him during those forty minutes. Everyone was grabbing him, trying to sit him up and shake him back to life.

I asked him, "Were you journeying into another world?"

Jorge did not answer me directly. He said, "In the animistic world view we feel that everything is sentient. Death is just another life-form. So I know what death is about and I'm not afraid of it."

I asked him what he experienced in the death state.

He told me that he felt like he was traveling without any goal toward some objective that he didn't know. But it all happened as if it had been seconds. In reality it had been more than forty minutes. He said that he had a hypothesis about what it is that we have to study about this other world that we don't know.

"What is the hypothesis?" I asked.

"The hypothesis is to look for, to search for, explanations that show that this other world is superimposed on this material world but in another dimension."

Mike then added, "Basically the shamanic explanation is that the whole physical world, everything in it, is an expression of the spiritual world. It is an overlay."

Then I remembered someone who actually experimented in the "overlay" on the frontier of life and death. As you might guess, there aren't too many of those types around. Oh, yes, there are many heroes. We read about them every day. But this was a person who moved into the twilight zone between life and death and then came back and reported just what he experienced. I was fortunate that this hero existed. I remembered when we met.

THE SHAMANS FROM THUNDERHOUSE

In early September of 1988, before returning to the United States, Judith and I left Paris motoring toward the east of France. We crossed the Rhine

River into Germany and then drove our small Renault across the southern German border into Switzerland. After a drive along Lake Constance and some searching, we managed to find the home of shamanic researchers Holger Kalweit and Amelie Schenk. I wanted to visit with Holger, who was the author of *Dreamtime & Inner Space: The World of the Shaman.*[1] Holger is an ethnopsychologist with degrees in both psychology and cultural anthropology. He studied shamanism in Hawaii, Mexico, and Tibet. He is also the founder of the Orpheus Project for the Study of Near-Death Phenomena in Germany and an active member of the German Transpersonal Association.

Holger Kalweit, in his travels, collected first-person accounts of tribal shamans from Oceania, Asia, and the Americas. In his book, and during our meetings with him, he described many of the shamans' paranormal experiences, as well as their rituals, out-of-body journeys, their use of sacred drugs, their descent into the underworld, healing with crystals, and other healing techniques.

Holger is married to Amelie Schenk, also an anthropologist, who worked with shamans in Tibet and India and shares his life. When we found Holger and Amelie, they were living in northern Switzerland near the German border in a small farmhouse near Lake Constance. The town they lived in was called Donnershaus (which translates to "thunderhouse," to the best of my Swiss-German ability).

A MAP OF THE MIDDLE WORLD

While I knew of Holger's visits with other shamans, I hadn't realized that he, too, had taken several shamanic journeys, using natural but quite potent herbs to alter his consciousness. To prepare himself for the experience, he had gathered certain plants and herbs that can be found growing nearly everywhere in Europe. He used these natural herbs in a manner that surprised me.

After preparing an ointment with these herbs, in a base of lard or other animal fat, he applied the ointment to certain areas of his body, such as the pulse points on the wrists or the back of the knee joints. He told me that this was always a dangerous undertaking, so his wife, Amelie, was always nearby monitoring his progress with the drug whenever he "tripped."

When he had achieved the effect he was seeking, he would know it, because he would feel a gradual numbness growing through his body, first

[1]Kalweit, Holger. *Dreamtime & Inner Space: The World of the Shaman.* Boston & London: Shambhala, 1988.

in his limbs and fingers and finally in his whole body. He would then lie down, with Amelie watching him carefully.

The first phase of his journey was marked by the vanishing of his sight. All during this time, even though he was blacking out visually, he would never lose consciousness. This *dark* phase was accompanied by a sense that he was physically moving, even though he was actually lying down. He would feel himself sometimes moving upward or downward. Sometimes it would seem to him that he was flying. He wouldn't see where or in which direction he was going, but he sensed and felt his motion.

I asked him how he knew he was flying. He told me that he could sense objects around him. He knew that they were there even though he couldn't see them. This experience lasted for maybe five minutes or less.

Next he went through a second phase, which I call the *tunnel* phase. Many people who have near-death experiences describe a similar experience. They see this as having first entered a tunnel and then moving toward a light that they see at the tunnel's end. Holger wouldn't necessarily see a light at the end of his tunnel. It could be more darkness awaiting him in some cases.

This movement through the tunnel led to the third phase, which Holger calls the *paradox* phase. In the beginning of this phase, he remembers himself and he begins to exercise some control over the journey. He literally "wills" that a vision occur. He commands what he will see next. With that control, whatever he wants to see appears to him suddenly. For example, if he wanted to see a tree or a village, it would just suddenly appear to him. He would see it the same way that we see the vision surrounding ourselves.

The perspective of his vision, that is, the viewpoint he would have, would appear totally normal to him. He could be walking along a road into the village or walking past the tree, or he could be flying over the village or tree. I asked him to give me more details about this "willing" and its connection to his vision of what was happening. He told me that what he saw was essentially the same as if he were viewing the scene with his normal waking consciousness. For example, right now if you look out at any scene before your eyes, you might see a particular object that you were looking for. That object would stand out in your vision, while the other objects around it would blur into indistinctness. If you were looking at a field of corn, you might not notice the grass surrounding it, for example.

Things would also appear to him in this phase of his journey to be completely normal, but after a while he would notice that something was missing or out of kilter. He once saw a person walking by who looked pretty okay except that the person didn't have a tongue. Another time he saw a tree, but when he looked down toward the bottom of the tree trunk, where the tree should have joined the ground, he saw that the tree had no roots. Another time he saw a person without a stomach. There would always be something wrong in this phase—something missing, as if one had wandered onto a weird Bosch painting.

Soon he would grow bored with this experience, and he would then enter the fourth phase. This phase appeared to him to be a different world. I call this phase the *trans-sensual* world. This phase is the most difficult one to describe. It transcends both space and time perception. For example, he would be able to see in several directions, up, down, right and left, all simultaneously. Yet, he wouldn't necessarily describe this as a visionary state. He said that he "saw" structures around him, objects that he enjoyed seeing; however, he could not describe them in terms of familiar geometric patterns, such as having lines and points or particular shapes.

This phase was also marked with the crossing of the usual senses, such as other shamans have described to me. He said that he could feel colors, or even smell a vision, perhaps hear a color or see an odor. This crossing over of the senses is, I believe, explainable in terms of the new model of consciousness I put forward in my recent paper.[2]

Then there was a fifth phase. After Holger described this to me, I decided that it should be called the *parallel universe* phase. He said that he would enter a world in which an apparent logic and order could be perceived; however, it wasn't the logic and order of this universe. Or there would be a principal difference in the physical matrix. There would be a different space-time continuum. There would be a sense of more space and less time, or less space and more time. This wasn't a mechanistic change in which things moved faster or slower.

Time-space would be different but only in a metaphoric sense. Time-space and causality are the ordering structures of this reality. But there are different matrices as well that can be experienced. To define these and put these into words is difficult. Holger felt that it wasn't possible. As he put it, the world he sensed was either "timey" or "spacy." There is some time and some space even in this imaginal realm or middle reality world. It is different in some characteristic way that is difficult to describe.

Holger's description had amazed me. Perhaps, without knowing it at the time, I was learning the territory of another world. I was also realizing that the connection between death and time also had another factor: parallel worlds. My sixth hypothesis, that shamans enter into parallel worlds, fit in

[2]See my paper in the *Journal of Theoretical Biology*, referenced in the bibliography.

nicely with the ninth, that shamans enter the death world to alter their perceptions in this world.

The death experience enabled one to cross over into another world, and the normal sense of time we carry with us from this world does become distorted. This element fit nicely with some of my earlier ideas about parallel universes and time travel. Let me explain.

Holger's first phase, the *flying in the dark phase*, could be explained as the beginning course of "soul physics." It is dark because the soul is beginning to move beyond light speed. It experiences motion very differently because it is moving faster than any light. Consequently, any light that it would emit would be dragged away in a wake similar to the wake around a boat moving through the water. The water waves move slower than the boat and consequently leave a wake behind. The water waves in front of the boat form into a bow "shockwave."

The second or *tunnel* phase is the soul's attempt to orientate itself to its own movement. When an object begins to move toward light speed, light coming into it begins to focus itself into a zone of light ahead of the moving observer. All light does this. It is something like what you experience when you walk into a rainfall. The rain appears to be coming from in front of you even though it is falling straight down. Near the speed of light, the incoming light forms itself into a kind of "light-tunnel." I would imagine that Holger's soul was now slowing down to just in excess of light speed and thus he was picking up on the tunnel experience.

In the third or *paradox* phase, the soul has managed to glimpse light waves, and it attempts to integrate these into an experience. Thus it has an experience of something normal but always something paradoxical. This probably has to do with the incompleteness of the picture received. Thus sometimes a person would appear without a mouth or other organ.

Holger's paradox phase may be explained by making an analogy. Normal waking consciousness is like tuning a radio to a single frequency and receiving information from a single station. I reasoned that Holger's experiences were equivalent to tuning the radio to a band of frequencies. Instead of experiencing a single reality, Holger was experiencing an overlap of different realities.

These different realities, like possibilities in quantum physics, can add together or be superimposed. This would result in an interference pattern or bleed-throughs from one reality into another.

In the fourth or *trans-sensual* phase, the soul again tries to integrate itself, but this time with the knowledge of the first three phases. Thus it is able to transcend the normal restrictions on space and time. Faster-than-light observers are able to make space appear as time and time appear as space. By "playing" in this dimensional crossover region, the soul finds itself able to witness in several spatial directions at once and to have sensory crossover as well.

In the fifth or *parallel world* phase, the soul is more direct in its action. Here it is able to visit other parallel realities and observe the "game rules" of whatever universe it happens to be visiting.

All of these phases and Holger's experiences in them made sense to me, particularly in the hologram model of the universe. Picture the universe as somehow frozen in space-time. It is like a gigantic four-dimensional (4D) film emulsion containing information from all parts of the universe. Like a real hologram, each part of the 4D hologram contains the whole. The emulsion has in it recorded waves of interference coming from all the universes.

Now just like a reference beam shining through a real hologram and producing images of the world, the soul is the reference beam for the 4D hologram. It has this ability because, not being material, it is able to move beyond the speed of light. It moves through the 4D hologram in different ways. Sometimes it is focused and at other times it is diffuse. Sometimes it moves through in one direction and then at other times it changes direction. Since it is beyond the speed of light, all of space-time is open to it. Thus it sees birth and death not as a beginning and an end but as, simply, parts of the hologram. Thus death was just another experience for the soul.

If Holger had actually died, his soul probably would have remained focused on one parallel world, the death realm. Then I remembered how I learned about the crossover to the death world.

THE EDGE OF HEAVEN

I had first learned about the crossover experience, between life and death to the other world, from Candace Lienhart, the American shamanka in Arizona.

I had asked Candace about her own death experience. She told me that she had been in a van with her fiance, Tom. Tom was driving the van when, unexpectedly, a tire came off the rim; when he attempted to pull over to the side of the road, the van struck a culvert and tipped over. Tom was killed and she was thrown out of the vehicle through the windshield and the van rolled over her.

She remembered experiencing herself as "sitting on the edge of heaven." And with that she felt herself moving through a tunnel, "like going down a giant slide." She felt herself travel back to the scene of the accident on the sound of her own voice.

Although her back was broken in six places and her face mangled, today Candace shows not a scar of her experience.

THE DEATH DREAM

Shortly after this dreadful experience, she had a dream in which she met with her fiance once again. In the dream she was in a canoe with Tom, who was holding her, wrapping her in a blanket, as she nestled in his arms with her back to his chest. Tom told her that she had to return. She then slipped out of the canoe as it tipped and felt herself falling downward through the water, which appeared to her as a stew or soup. As she descended, the soup became more and more dense, and it became more and more painful. As the pain increased, she felt herself coming back into her pain-racked body, and she knew then that she didn't have the choice to return to Tom anymore.

As she was returning to the earth level, she saw many things that stuck with her to this day. For weeks after, she wondered why she had been brought back. When she looked at people, she could see things about them that we normally can't. She calls it seeing them in 3D. There were the dimensions of the front facade, the middle facade, and the background spirit. She saw everyone in three layers, as if there were an overlapping merger of images. She found this very painful to witness. The near-death experience had wiped away her natural defenses against this type of seeing. She felt that she had to observe the layers of the people she witnessed to see who they really were.

After her ordeal, her perception shifted, cushioning her out of "seeing" indiscriminately. Today she still has the ability to "see," and she is well known in the western United States and Hawaii for her visionary and healing abilities.

SHAMANS, DEATH, AND THE JELL-O MIND

Candace had reminded me that in her world vision, the usual assumptions about time and space do not apply. We are all living our lives as if they had beginning points, our births, and ending points, our deaths. But from a new physics pinnacle, everything is connected. Life flows between points and doesn't simply begin and end with birth and death.

Candace explained that, in her experience, most human beings will say you have only one death experience. This experience shocks your system to a "Jell-O state" where you have access to all areas of *what is*.

"What is a Jell-O state?" I asked.

She explained that a Jell-O state is one with no boundaries. Your consciousness is free-floating without what she called "alarm system boundary layers" in between. It is full-on-knowingness, all at one time. She saw it as "full consciousness."

She said, "The big quest is to full-on-knowingness in the physical realm. But what people forget is that the density of matter, itself, requires a slowing down of consciousness in order for it to manifest on this plane as matter."

Candace told me that her experience of shamanism was a continuous death and rebirth sequence, on every level. To understand the actual aliveness of this plane of existence, you must embrace the conversion of everything continuously changing. Death is an ally for the shaman. He or she must experience death several times.

I asked her to give me an outline of death based on her experiences.

She told me that in one of the death experiences that she had, she basically felt and experienced herself as being all that there is. She said, "I was like a dot of light that was connected to all the dots of light, and I could be as big or as small as I chose to be."

Her experience of people who have died, however, is that when they die they go to where they think they will. This opens up another realm of what she calls psychic rescue service. This was the same idea that the pendragon had told me about earlier. There are people who get caught in their limitations when they die. They actually go where they believe they will go.

Then she told me something that confirmed what I had been thinking about the nonpresence of time in the death state. She said, "Some of the vagabond-poltergeist stuff we hear about and witness isn't happening from a big intelligent point of view. It's being caused by a noisy trapped spook who is tired of being where he is, but since there is no time where he is, he can be there for ten centuries and not know that any difference has taken place."

THE MIDDLE REALM

I then told her about my visit with the shamans in Britain, who work within a system they call wyrd. They told me that there are two realms. There is a middle world realm of death and there is a spirit world realm of death. I wanted to know if there were any other realms and what the difference was between these.

She explained that the middle world is a realm where the individual has not chosen to completely release what was happening here, on this plane of existence. They still have investments that they want to look over. They don't have freedom. They have more density than the spirit realm but less than the physical realm.

That meant that beyond matter and energy there was a physics of consciousness—pure consciousness. I began to suspect that these other realms were indeed questions of consciousness density. That meant that consciousness itself was capable of forming itself into patterns outside the material world. One of those patterns is what we call the soul.

Candace explained that people caught in the middle denser realm have not awakened to their true unlimited nature. She called it their "infinity." She said, "They haven't really embraced that infinity. If they embrace the infinity, then they are free. They can be anywhere, go anywhere, manifest as anything according to what their purpose is."

I began to think that these other realms were possibly parallel universes, but on a very different level of reality. Instead of being physical realities, these were in the realm of mental realities. Possibly the laws of quantum physics also applied there, as they did in the physical realm. If so, there had to be more than just two or three realms.

Candace explained that, from her experience in these "other worlds," there were many, many layers in between. She said these worlds were like the colors blending in a rainbow, in contrast to my earlier speculations that they were clearly separated. They were to her as thin as baklava. The amount of spatial occupancy that they have is finer and finer and finer. She said, "When you move into it and you are experiencing it, it may seem a million miles wide but it isn't. It is actually very thin."

Candace then shifted her attention. She began to tell me about the tunnel that people say they go through when they die. She believes that this tunnel is a spiral of their experiential existence. They are moving through their own infinite spiral back to the point of creation, and they are going through what we call Karma and memory sequences.

She explained that all Karma is unfinished business, leftovers. That is all that it is. When you have gone through that spiral and routed out all of your leftovers, you are finished with life on the material plane of existence. She said, "Fred, in every reincarnation movie, if you haven't finished your homework in one past lifetime, it will show up on your desk in this lifetime."

As she saw it, there is really no need to go through and contact all of the past lifetimes, because all of those lifetimes will be with you in this time frame. If you deal with what you have got in this now-time and you move to your point of peace with all things, balance within your system, then you find that all your Karma has cleared up, and all your past life sequences are fulfilled.

The Chumash medicine people, the shamans from Brighton, the American shamanka, Candace and nearly every other account of death I have read dealing with the shamanic world, all agreed on this point of reincarnation. You choose to come back to complete whatever you felt incomplete about.

Clearly death was not the end of consciousness for these shamans. Since they were able to access worlds in the death state, while they were standing in a healing ritual with a patient, shamans had to have learned to master space and time. At least they had to know how to do this with their minds.

Again the example of the 4D-hologram universe came to mind. Our brains are made of matter, but our minds are made of a nonmaterial "intent." This intent acts like a laser beam. It shines through our material substance and gleans information. This information is recorded in our whole body. When we die or go through a near-death experience, this "intent" remains. Only now the whole universe becomes the hologram, not just the brain, muscles, and nervous system that make up our bodies.

Shamans have dealt with death and near-death, so they know how to move their intent outside of their physical bodies. As holographic beams, they are able to enter the universe as a whole, travel to other parts of it, and even become part of the consciousness field of animals. Death is not an end for them because they have, in a sense, already died many times.

I was convinced that each of us also has this ability. It is contained within the five extra senses that the Chumash medicine people told me about. It is the ability to reach into the imaginal realm and glean information from it. Since the imaginal realm was that part of the universe that was faster than light, it was that realm that shamans went to. In it the normal time sense, dictated to us by having material bodies that move slower than light, no longer holds. We travel through physical time in the imaginal realm as if it were space. It no longer has a hold on us. The only time sense that exists there is "mythic" time. The only time sense we have in the faster-than-light realm is that which we carry with us through memory. That memory is held as records in the 4D hologram. And being faster-than-light souls or beams, we are able to access that information in many different ways and patterns, without any barriers due to physical time.

I still needed to follow up on this lead. If quantum physics, parallel worlds, and death or near-death were related, then, as I suspected, it was finally possible to understand how shamans worked. The secret was contained in their ability to access a form of awareness that I call shamanic consciousness, and in that state to walk in the mythic world of destined reality—a far cry from the indeterminism of the physical realm.

II

Shamanic Consciousness

Miguel has completed his own ayahuasca journey. He now recognizes that death is his ally. He has awakened to a new vision of himself and his work. He knows that as a sociologist returning to the academic world in Lima, he has a story to tell. It is no longer just an academic curiosity that he feels. He has lived the life of a shaman-initiate. He knows that beyond chronological time there is another world, a mythic world that contains him as a small part. He begins to sense his own power as a mythic being. He begins to see that his journey to Iquitos was more than just objective facts. He sees that the events of his life told more than just a factual story.

THE SLEEPER AWAKENS

And I, too, was having this realization. Scenes of mythic revelation were flashing in my mind. I remembered the fly on my foot in the temple in Nepal and the vibration of Don Solon. What were these events? They were reminders, intrusions into the temporal world from the mythic world. They were stories of an awakening giant: Mythos. That giant has no physical size. It is as big as the planet. It is as small as the three-pound universe we call the brain that exists in our head. These stories were the elements of the overmind, the mythic reality without which, in shamanic reality, nothing physical would manifest. But these elements came not just from the individual overmind. They came from the overmind of the universe itself.

The overmind was like the aleph in the Qabala. It had no energy but it contained all possibilities. It was pure potential. To see in this reality of pure potential was to walk the shaman's path. This path was a creation of a new least-action path. It demanded action. It demanded that the sleeper awaken from the automatic pilot it had left in control of the universe.

To grasp shamanic consciousness, to see what they see, we must be able to "think" and "reason" in two very different worlds or realms. These worlds overlap, but the overlap is fleeting. Only when special circumstances arise does the overlap persist. For a few individuals, the overlap of the worlds persists for a few moments. For the shaman, the overlap of the two worlds is the world he or she walks in.

But to even begin to see shamanically one had to believe that the other, mythic world existed. I remembered the famous story of Captain Cook visiting the Hawaiian Islands. When Cook and crew came ashore, they had to leave their large ship floating several hundreds of feet distant from the shallow shoreline. Cook pointed the ship out to the native Hawaiians. They didn't see it. In fact, even when Cook rowed back out to the sailing ship with some of the natives, those natives still failed to see it. They didn't see it until they nearly had their noses pressing against its sides.

Why? Because they didn't believe that any ships could be built like the giant sailing ship. They had never seen a ship that big before, and so why believe in something that you have never experienced in your life?

A single experience will convince anyone of the reality of something. But to have that experience, unguided, without any Captain Cook to point us down our noses, is difficult, if not impossible. The reason is simple to state: You must know what you are looking for before you *can* see it.

Even Einstein said that it was the theory that tells us what to look for in the experiment. Without a theory, data would fly by unnoticed. Our everyday lives consist of many overlaps into shamanic consciousness. The problem is we normally don't know what to look for when this happens, so we simply fail to see them. We are like the natives visited by Captain Cook. Until we either experience a shamanic event or someone takes us by the hand and rubs our nose in it, we won't see it. And not seeing it, we won't even believe it.

THE "REAL" REALITY

An old saying is "Seeing is believing." But my quantum physics, together with my shamanic experiences, was telling me that we had it reversed:

"Believing is seeing." We will see it when we believe it. In some sense, each of us "creates" the reality we see out there from our beliefs. At the moment of perception, the moment that a record of an event is "stored" in the memory of the observer, the object, being observed, manifests.

But what does "create" mean? Let me explain using an analogy of a camera coming to a focus. An object appears as a result of observation in much the same way a blurred image suddenly jumps into clarity when looked at through a lens that has just been focused. Out of the blur comes a sharpness.

But unlike the film in a camera that just records a clear image from a blur of possibilities through the action of a focused lens, the recording of an image of an object in the mind of the observer *creates* a clear object in the world. The action "out there"—the sudden appearance of the object—is simultaneous with the observation of the object. It is a mind-matter link. It is the action of consciousness.

This is still an unsolved mystery in physics. We still don't know how this action takes place. We only know that it must take place. Every attempt to figure out how this occurs, to in some way observe it, to experimentally test it, has failed. This simple act, the recognition that a recognition of an objective quality is linked to a subjective perception, appears to lie on the boundary between reality and fantasy itself.

I had felt that boundary when I gave up my university post at San Diego State University in 1977. It was as if I had suddenly awakened. I walked the corridors of my university and looked in at my colleagues. They seemed to be dream-walking, going through the paces of their academic existences, waiting for retirement. Everyone there seemed to be moving as if in slow motion. I felt invisible. If I moved like they did, they wouldn't see me.

I had felt a yearning inside of me to escape the traps of physical time. It appeared to me that my colleagues were doing their jobs like the slaves in Plato's cave. These slaves, having spent all of their lives chained to the wall of the cave, could only see their shadows when a light shined on them. They mistook their shadows for themselves.

Perhaps the physical universe is like Plato's cave. We are only witnessing the movements of shadows, illusions. The *real* reality is in us, but we can only sense it through our intuition. That reality appears to us in the dreams that Jacques and I were talking about.

My colleagues had seemed more dead to me than alive. I became afraid. Was this to happen to me? Was I really seeing what was out there? Or were those just feelings trapped inside of me?

THE DROP OF CONSCIOUSNESS

And now in Lima, watching the movie's final scenes, I had a strong feeling inside of me. I saw that my body was a trap of consciousness. This idea of consciousness being trapped is important. It is in a trap of its own making. It is a stable system of energy. It is self-consistent and self-continuing. It keeps seeing only itself apart from the rest of the field or ocean of consciousness. Yet it is a drop of the ocean that, because of the working of surface tension forces, tends to hold itself as a drop, separated from other drops, separated from the ocean of consciousness.

We become drops floating in the atmosphere, mostly unaware of the ocean that lies below us and the gravity that will eventually pull us back down into it. In order to form a drop, a microdrop of water vapor that will escape gravity and rise from the ocean, it must separate by forming a surface that holds the drop intact, separate from the other drops.

The rest of the world appears, then, to be separated from ourselves. The drops are suspended in air, until they fall back into the ocean. The fall into the ocean is our own surrender to death, where we merge once again with the ocean of consciousness.

Of course, as we fall, which is our path through life, we pick up all kinds of stuff in the air. Dirt particles, dust, atomic or molecular debris of one kind or another. This works into the drop, giving it structure or character. This character usually is embedded on the surface of the drop. This surface is our ego, our mental shield protecting us from each other. Our problem, our addiction, is that we believe in the stuff we pick up, the molecular debris, but fail to identify with the original substance of the drop—the bit of consciousness from the great ocean that we came from and to which we all shall return. When we do, all character is eventually dissolved away. Any micropieces of rock we picked up during the fall dissolve away eventually.

That dissolution takes time even when the drop returns to the ocean. Thus when we die we still carry, for a time, a sense of our material life, our material addiction, our egos. But eventually the drop merges back into the ocean, waiting for the sun to evaporate it back into the atmosphere over and over again. Whatever material, whatever dirt, we fell into the ocean with is, of course, dissolved in the ocean, but it remains there. It doesn't go away. Thus when we evaporate into drops again, some of those particles of dirt are carried with us. This is our Karma, our habits from previous life experiences. We carry those into the atmosphere of daily life, and unless we learn how to free ourselves from these habits, they become part of our surface armor, our egos.

Just as water-drops are part of the ocean, individual consciousnesses are part of the one consciousness. In this way, we are one, brothers and sisters. There is no difference. The self-trapping, the formation of the water-drop surface, makes it appear as if we are not one, but many.

I was lucky. Although I didn't fully know what to look for when I entered the shamanic reality, I was trained by the shamans of the world. They were telling me what to look for. I was realizing that what there was to look for was the same as I had believed when I was a child. The five senses of the imagination were all that I needed to heal myself, to see into the future, to see into the past, and to walk the shaman's path. The only sacrifice I had to endure was giving up my beliefs that this could not happen.

Later that evening I realized that I was beginning to have a theory. My third hypothesis told me that shamans perceive reality in a state of altered consciousness. My fourth hypothesis said that shamans use any device to alter a patient's belief about reality. These were the clues to my theory. Why did shamans use devices, any devices, often just simple conjuring tricks? Of course, the answer was to change the patient's belief structure. This would create a belief in fantasy in the patient. But why fantasy? Presumably fantasy altered reality.

Now it was clear. There was a shamanic physics, and the fundamental law that governed it was found in the shaman's ability to alter the way he or she observed the world. I knew that the mind of the observer entered into the world of quantum physics and therefore had to affect the objective world of matter and energy in atomic experiments. But now I realized that shamans also had the ability to alter physical reality when they entered a shamanic state of consciousness.

But how did they do it? I began to piece together my thoughts. Their power was the ability to change least-action paths. They did this, not by changing the beginning and the end of those paths, but by changing what happens in between.

Let me explain this. Since objects follow all possible paths in moving from A to B, what determines where an object will be in between A and B? The answer was: The observer determines what happens in between. When the observer looks, the path of the object comes into existence, much like the camera image I explained earlier. This much, you'll recall, Werner Heisenberg taught us more than a half-century ago. Somehow the very act of observation "creates" the observed.

Now when a physicist creates a least-action path, he has no vested interest in what the outcome will be other than having the outcome confirm his theory. Consequently, he would be surprised if the outcome deviated from statistical predictions. So he rarely sees any mid-least-action paths deviating from an expected trajectory. For example, in a bubble chamber, he sees the path that a subatomic particle takes to be what he would expect it to be—a straight line. Now the probability is very high that a track in the bubble chamber follows a straight line. And thus, physicists expect to see that happen, and it does.

Shamans do not deal with probabilities. They are not surprised when things do not follow straight lines. And now I knew why. Shamans were able to change the least-action pathways by entering into a mythic world.

This thought seemed so simple. Somehow they were able to use the data of the physical world and the images of the fantasy world to create changes that would seem miraculous in the physical world. Shamans were able to live in the overlap of the two worlds. In a very real sense, shamans used the world of imagination to alter God's crap game. This was something that physicists would never do because they were never trained to believe that it was possible.

So if shamans, and presumably other human beings, were able to alter the probability game of physical reality by bringing in fantasy, there should be some evidence that this was true. How would this be observed? Probably by observing the extraordinary sensitivities of shamans in action—in other words, observing them changing mid-least-action paths. What would that look like? It would appear as an ability to know, discover, or change a physical system through a means that seemed to violate the laws of physics. Now with the theory in mind, what were the experimental data?

SHAMANIC SENSITIVITY

I then remembered my meeting with Richard and Chris, the shamans from Brighton. Chris had told me that Richard still had not lost his shamanic sensitivity. I knew this was true because of the way Richard first told me about the Qabala. He spoke to me as if we had been conversing about it for the past few hours, as if he knew I had an interest in it, and yet I hadn't even mentioned it.

Chris sat down with us, holding a cup of tea in his lap. He was listening to Richard, and then he told me a story about Richard's sensitivity to the planet. One day Richard and he were out in the woods. They had taken an academic friend with them. They were deep in the woods and then Richard said to the academic, "Now where's north?" Chris had turned around and said "It's where that hawthorn bush is." And Richard said, "No, no, I think it's just a tiny shade to the left, just about the left portion of the hawthorn bush." Then Richard and Chris agreed that they were pointing to true north.

Being a typical academic, their friend was meanwhile checking it with his instruments. And he found that they were absolutely spot on. The academic couldn't believe it.

I laughed at myself listening to this story. I particularly found this amusing because I knew that academic training often had me relying on my external instruments far more than the ones I was born with. Perhaps I had lost my sensitivity to my planet when I gained my understanding of a compass.

But how was I to regain it? I then remembered my meeting with Holger Kalweit and Amelie Schenk, the shamanic researchers I met in Switzerland.

Holger and Amelie told me that shamans often use shock, such as a physical jolt, to alter a person's consciousness. We often use words to achieve a similar thing. One only has to think of any military training experience to recall how words can alter one's consciousness. A crack military team performs as a unit. They are sensitive to each other in ways that cannot be explained by logic. They must believe in something bigger than their individual selves.

Amelie described an initiation ceremony that she observed in Ladakh, Tibet, performed on an epileptic woman who desired to become a shaman. To shock her out of her normal consciousness state, she was chased around a fire by her shaman-initiator, who wielded a sword and drum. He literally chased her around the fire threatening her with her own life.

Amelie was photographing her while she ran around the fire. At first the woman being initiated was petrified with fear, but then, as the ceremony progressed, she actually began to smile at Amelie, indicating that her fear had been transcended and she had gone into another state of consciousness.

The shaman-initiator then performed a series of tests to see if she was capable of using shamanic power. Before this ceremony, any attempt by her to play drums or rattles was marked by extreme ineptness and clumsiness. But after this chase ceremony and her entering into a trance state, she became perfectly adept at using these instruments.

There was a test to see if she had developed a shamanic state of consciousness. Before she went into trance, the shaman performed a little ruse. He took a certain number of small pebbles in his hand, and not showing them to her, he asked her to divine the number of stones. Before trance had been achieved, she was unsuccessful in guessing the number. But after achieving her trance state she was always successful, no matter how many trials she was put through, and "knew" how many stones there were.

She was also able to find buried white silk honor scarves, used by the

Tibetans as gifts to guests who visited them, that were hidden from her. In another test a burning candle was placed next to her forearm skin. If she had achieved trance, she would not be burned. After applying the candle flame to her arm, her skin showed no damage. Another test used a white-hot knife that had been left in the fire for a while. They touched the knife to her exposed tongue. Again she was not harmed while she remained in the trance state. All of these were tests of her shamanic power. And amazingly enough, after she had achieved her shamanic power her epilepsy also disappeared.

Holger then told me about another epileptic, an Englishman named Adrian Boshier, who decided to become a shaman and undergo initiation in Africa.[1] After achieving shamanhood, his epilepsy was healed. However, he was warned never to bathe in salt water or else his epilepsy would return. One day he forgot this and took a bath in the sea and drowned. Actually, according to Lyall Watson's account in *Lightning Bird*, Boshier didn't really drown, he just sank into the Indian Ocean, but when he was brought back to the surface, he was dead.

In Australia, epileptics are considered to have godlike powers. If an epileptic wanders into a tribal village, he is offered a home and food and generally honored as a holy man. We now know, through the monitoring of slow brain wave patterns, that an epileptic's fits show a drastically altered brain rhythm.

As we wondered about the connection between brain rhythms and consciousness states, Holger offered a theory. Holger had some ideas about the relationship between trance and mind. He had observed many shamanic ceremonies using various percussion instruments—drums, bells, and rattles. He noticed that there was a definite correlation between the vibrational frequency used by shamans to go into trance and the theta rhythms in the brain. He is convinced that the trance state is a theta state. The frequency used by the drummers is the same as a theta frequency.

Holger also witnessed shamans who could bend a metal sword merely by touching it, much as Uri Geller had done in Europe and the United States. In every case that Holger and Amelie witnessed, the shaman had always undergone some experience that shifted him or her into an altered consciousness state.

If Holger was correct, then the use of vibrational tools such as drums and bells could be the barbells we needed to build up our psychic "muscles." Just as a bodybuilder must believe in what he or she is doing to build the

[1]A description of Boshier's adventure in shamanism is described in Watson, Lyall. *Lightning Bird*. New York: Simon & Schuster, 1982.

body of his or her dream, and a karate blackbelt must believe in his or her ability to "tune" to his or her opponent, we need to believe in the imaginal realm—the realm of the myth—if we are to touch it. We must learn to do this in spite of our technology, which seems to be acting in a manner contrary to that wish.

It was clear to me that when shamans entered into trance consciousness they were able to alter physical reality. It was also true that shamans who had not been trained in physics would not have any reason to not believe in their seemingly magical powers.

Belief must have a lot to do with what we see in the world. Several people have reported seeing other realities. Were these just fantasies? Perhaps we all could see other realities if we believed in them. The ley line researchers, Charla and Paul Devereux, had told me about the assorted anomalous phenomena such as the time-slip sightings that occurred in Oxfordshire. Throughout our history such sightings have been reported. Even today we are faced with other reports, such as UFO sightings, and even reports of people being abducted aboard UFOs. Whatever the "real" truth is behind these phenomena, I realized that to grasp shamanic consciousness, one must seek out the unusual. One must leave the path of habitual behavior. But this wasn't easy to do. Our bodies are run by habitual consciousness.

THE MATERIALIZATION OF CONSCIOUSNESS

I then thought again of my visit with Jacques and Jose in Tarapoto. We were talking, and Jacques began to say something when suddenly he stopped. A strange look came over his face. After a while, he looked at me and said, "Just now, I had a physical sensation. The Indians don't speak of mental sensations. They speak of the body. I realized that my body is a materialization of unconsciousness. My body is not a symbolism of my unconsciousness. It is my unconsciousness."

I became excited. I had made arrangements to speak about this in connection with how addiction and the unconscious were related, at a conference on drug addiction in the United States when I returned from Peru. I had felt the same realization. The body is the materialization of the undermind. In a sense, the body itself is an "addiction of consciousness." But so is the whole universe. Addiction and matter are one and the same. But granting me this for the moment, let me try to explain this rather preposterous statement.

First we need to ask: How did consciousness materialize and become the body? According to Buddhist thought, matter arises from desire. Any desire seeks fulfillment through repeated experience. That repetition appears as a frequency, a vibrational pattern. And from quantum physics we know that all matter has a vibrational pattern.

Our bodies become addicted to material substances. In the same way, our consciousness becomes addicted to material form. It is an addiction or, perhaps less strongly put, a desire to have something rather than nothing. It starts out as only a thought, say, in the mind of God or the universe, or a higher power, for this to happen. But then the desire arises for something rather than nothing.

It is also a sacrifice. Everything at the overmind level is mythic. It is here where *will* and *intent* live together. It is here in the overmind that nothing exists. But in the undermind, the body-mind, the unconscious, everything exists. It exists at a price. That price is becoming unconscious and following least-action paths. In other words, becoming mechanical in action. It is there, but it lacks heart and mind. But life involves both the overmind and the undermind. Life is myth and survival—this life that we are living right now.

All theory is myth. In fact, even the so-called physical truth of the universe is a myth. It has no more reality than any other myth. And just as we do in any other myth, we see a truth in it. But at some point, a very deep one in modern physics albeit, the myth begins to unravel. We find logical inconsistencies. So we try to rationalize the myth, turn it into a deeper truth.

Jacques, investigating ayahuasca for several years, told me that he was happy to hear me say this. He had felt very alone when he tried to explain this to his colleagues because he did not have the knowledge to back up this intuition. He said, "I usually find myself confronting rationalists, getting nowhere. In my article,[2] I wrote that I felt close to the ideas of fundamental physics. This kind of language that you speak is not strange or odd for me. I understand it at the intuitive level. I can't understand it at the mathematical level. I understand that the conceptual framework is the same as I have discovered using ayahuasca."

Jacques's face faded from my consciousness. Yes, ayahuasca was my teacher, too. I realized now that it had taught me how to enter into a shamanic

[2] Mabit, Jacques Michel. "L'Hallucination par l'Ayahuasca Chez les Guérisseurs de la Haute-Amazonie Peruvienne (Tarapoto)." *Institut Français d'Etudes Andines.* (IFEA, Casilla 278, Lima 18.) France: Document de Travail, 1/1988.

state of awareness. That awareness still persists today, even though my journey has ended.

I looked up at the screen. The movie was coming to an end. Miguel visits Meliton again before returning to Lima. He is more aware of the crack between the worlds than he ever was before. He realizes that his journey, at least for a while, is over. Meliton tells him that he was on a journey into his own unconscious mind.

I remembered all of my experiences with the Peruvian shamans. I remembered how I had felt when I was to take my last session with Don Solon and with ayahuasca. In many ways it was the most powerful session I went through. I actually had a conversation with myself in the future and myself in the past. It was as if the three worlds of time—past, present, and future—had overlapped. The overlap of the worlds occurred within me.

ACT CLEANING

I was waiting in the jungle hut for the return of Jorge and Don Solon from Iquitos. At around eight that night they finally arrived with Ted and two new people, and without much ado we began what was to be my last and fifth experience with ayahuasca. I didn't know it then, but this was to be the most profound of all of my experiences.

The newcomers were two students from Yale doing graduate studies on medicinal plants of the Amazon jungle. Their names were Trisha and Carlos.

The ceremony was a repeat of what had happened to me two nights ago. My visions were less coherent, however. I did see a vision which seemed to be of a sun god. I remembered that the Incas had come to Peru and had established that their king was a god of the sun. But the voices spoke to me once again and I felt extremely ill. I was suffering more than I had expected and was saying to myself, "How much more of this can I endure?" I asked myself, "Why am I suffering so much?" Then I heard the teacher speak. He was myself much wiser and older. He was my future.

He said, "Clean up your act, Fred."

Then something very strange happened. I became three persons. There was myself in the present. He just listened to the conversation. And there was myself in the future, the teacher. And there was myself in the past, the child who wondered about the illness of the body in the present.

The child asked, "But why do I have such a terrible amount of stomach acid?"

The teacher answered, "Do you remember your childhood? You were always in a hurry. You never wanted to pause; you always wanted to get on to the next experience. You ate hurriedly. And so you put food into your stomach so quickly that you hardly tasted it. This rush of food into your stomach caused it to produce stomach acid for digestion. Your stomach, like a willing servant, did just what it had to do, produce more acid. You have had more than fifty years of eating experience, and you have trained your stomach to do this. Now you don't eat as much as you did. Your metabolism has slowed down, but your stomach still acts as it was trained to do. Thus too much stomach acid."

"What can I do to heal myself?"

"Slow down. Be conscious of every bite. You must retrain your stomach. If you do, your condition will improve."

"Why am I afraid? Where does this fear come from?"

"It is a fear that you never will have enough. It is a fear that you will not be accepted or loved."

My conversation with my "selves" continued throughout the night. After the ceremony, I left the large room and went to my bed. In the privacy of my own room, I wanted to process what I had learned. I realized that the vine had opened up my unconscious mind and that all of these thoughts were just those that I had suppressed over the years. I realized insights into my family. I saw that I had a resonance with each of my children, and that if I was honest with myself, I would be able to see them more clearly.

The healing was working. Ayahuasca and the songs were a purge. The visions of the future that I had sought were secondary. I had to work on myself, be honest with myself, and thereby clean up my act. The importance of the shamanic world was not my discovery of how it worked objectively, but how it worked subjectively. It revealed my own true nature.

This time I welcomed the tobacco smoke when Don Solon blew the smoky wind over my body. I saw that tobacco in combination with the ayahuasca was capable of healing me. Not that I am suggesting that everyone smoke. That privilege is abused, and when tobacco is used as we do, it becomes a poison. But used correctly, tobacco can heal not harm.

My mind drifted and floated. I saw pink dolphins swimming in the Amazon River. Why were they there? Now I know, because I saw them in the film. I saw that the Amazon was the middle of the earth, an umbilical cord connecting the land with the sea. It stretched for thousands of miles, and nearly all of those miles right along the waist of the mother.

I noticed that, in the night, my dark-adapted eyes were also experiencing hallucinations. Two of the dogs had come into my room, and they were scratching, shaking the room as Don Solon had during the ceremonies. I reached out to push one of the dogs away from the bed, and heard a loud noise. The dog had changed. I had pushed my tape recorder off the night table I had placed near my bed! I laughed at myself.

With my mind filled, and the hypnagogic images racing through my illuminated mind, I fell asleep.

When I awoke, the birds were talking and it was eight in the morning. I felt quite well and my mind was filled with thoughts about the previous nights. Before I had retired, I told Ted that this last experience was the most painful I had endured. He told me that that night was a special one for me. Don Solon intended it as teaching. He wanted me to see how my personality was affecting my body and my health. Arising this morning I realized that I wasn't intended to have visions last night; Don Solon was directing the spirits of the ayahuasca to illuminate my self to myself.

I remembered that my two most sensitive areas, my stomach and my head, were both filled with fire. I suffered a splitting headache. I felt that all of my sinuses were throbbing and screaming. And at the same time my stomach felt as if I had swallowed molten lead. My journey was a body exploration, and I saw that my mind-personality was literally killing my body.

When Don Solon brought me to his feet, and blew the tobacco wind, my headache became even worse. Don Solon knew what I was feeling. He put his lips to my head and he sucked my headache away. After each application of his lips to my brow and to the top of my head, he spit out whatever he took in. Amazingly, my head stopped aching immediately when he did this. I had never before experienced such a rapid healing. It took only seconds.

Later Jorge had me lie down and he worked on my stomach, soothing me, and gradually, not as quickly as Don Solon's cure, the fire down below subsided.

I learned that my body is a reflection of my behavior. My stomach is a speed indicator. Any pain there means slow down, take it easy. When it hurts it tells me that I have overpassed myself. I should have learned this a long time ago. When I went through puberty, I began to have a severe stammering problem. One of my teachers said that my mind was working too fast, faster than my speaking mechanism could handle.

Meanwhile, while I was going through this ordeal, Ted was apparently having a ball. He experienced nothing at all of the problems I had. He was enjoying every moment of insight he got. I was amazed that he and I had gone through this together with very different experiences.

I soon got up, had a wash, and joined everyone at breakfast. As usual, we had rice and fried bananas. I decided to ask Don Solon about my experience.

He told me that on Friday nights he had to be especially careful because there were many *brujarias* (sorcerer spirits) around. "The other nights are more tranquil. It totally depends on the maestro's command of the situation."

"Do the spirits change if someone gets ill, like I did?"

"Sometimes when a person gets sick, it means that a *brujo* has gotten through and is using his power. Sometimes he does this to see if he has more power than the shaman."

"Was that what happened to me last night?"

"No. You should have your stomach checked when you return to the United States. The ayahuasca would not have bothered you if your stomach was okay. I knew your head was hurting. I sucked the ayahuasca from your head and that made the pain go away."

Ted then told me that he, too, experienced a headache last night and that Don Solon had done to him what he did to me: he sucked the pain out of his head.

I then asked Don Solon about his clouded right eye. He told me that a brujo had caused that to happen. He had seen several doctors and they all said it was not a cataract. To this day, he hasn't been able to see out of that eye.

One day he went to another shaman to see if he could heal the eye. The shaman used the sucking technique and applied his mouth to the afflicted eye. He then spit out small daggers of sharp wood from his mouth. Now these daggers or darts of wood were not in the eye. They manifested as material forms of the witchcraft that had been performed on him. The shaman told Don Solon that he had never seen this before. Don Solon believes that the brujo indeed used wooden splints in his spells.

Soon Arturo, the man who set up our stay at the jungle retreat, appeared and told us that we had to return to Iquitos that morning, and the boat was now ready. I had packed my things earlier, so I went to my room, got my things, gave some gifts to the people we had stayed with, and soon we were all on the boat making our way upriver to Iquitos.

On the way I talked with Trisha. I had noticed that she had gone through the ceremony without incident. I was surprised and asked her what she had experienced. She was quite a lovely woman in her twenties. She said that Don Solon had given her a very small dose and that she had some visual experiences but had no stomach discomfort. Jorge, on the other hand, threw up last night. Later I found out that he was a bit careless the night before and had taken too many *cervezas*. I was also surprised at that. Trisha was in the jungle to help native people harvest jungle plants, telling them which plants would be useful as sources of medicine.

As she was talking I had a sudden premonition that possibly I wouldn't be going to Tarapoto that day. I asked her what she was planning to do in Iquitos. She told me about a birthday party that evening for Carlos and said that I was invited.

By noon we arrived at the Iquitos riverfront. I said good-bye to Trisha and Carlos. Jorge, Ted, Don Solon, and I were on our way to the airport to fly back to Tarapoto. Don Solon and Jorge were planning to provide at least two more ceremonies for us when we reached Tarapoto. However, I did not look forward to this. I wondered how I would be able to endure it. But as fate would have it, I didn't have to worry.

CANCELADO

We reached the airport, and were told that the arriving plane from Lima had had trouble in the morning getting its nose landing gear down. Fortunately no one was injured. Consequently a team of salvage experts had arrived from Lima and were clearing the crippled plane from the runway. That took hours and delayed any flights from landing. Our plane was due in by 1:30 P.M. but was now delayed until 4:00. We waited at the airport.

At 3:45 P.M. the announcement over the loudspeaker told us that all flights out of Iquitos that day had been canceled. We had to remain in Iquitos until Monday. I felt relief. Our driver had been waiting for us, just in case, so we all made our way back into town. We dropped Don Solon at his home, and then Jorge took Ted and me to the Hostel Amazonas on the Plaza de Armas, in the center of town. It was nearly five and by this time I was exhausted and still feeling the effects of the night before.

The hostel room appeared to me to be a great luxury. I had running hot water, an air conditioner, and a bed to sleep in without mosquito netting. It felt better to me then than a luxury suite at the Beverly Hills Hotel.

I agreed to meet Ted for dinner around eight, and after a wonderful shower and a shave, I lay down on the bed and took stock of my adventure.

How can I explain what I saw and what had happened? You can always say coincidence. Or it was just stress. It is easy to dismiss experience once it has happened. But when Don Solon sucked on my head, my intense headache just vanished. Usually my headaches subside after a while. After the first hour, having taken an aspirin or sinus medication, the pain begins to slowly dissipate. Never, with one other exception of the experience I had with the Indian shaman in Ojai,[3] have I had a headache vanish, bango, just like that.

THE WHOLE-MIND ILLUMINATION

I believe now there is an explanation, or at least an indication, of how I was healed. But to grasp how my body was healed, I needed to consider what I now believed was happening in my mind. On ayahuasca, my whole mind, including my unconscious mind, was illuminated. As I've described, you can think of consciousness as a well-collimated searchlight or, if you wish, a laser beam. Through life's experiences, each of us has learned how to use that searchlight. We use it, we aim it, allowing it to illuminate various parts of our cortex, in the same way that a searchlight scans the sky looking for airplanes in a typical World War II movie.

For example, I am about to cross a road. My searchlight scans the visual cortex and the aural cortex, paying close attention to any input data from my eyes and ears. My feet cortex is also scanned to make sure that the road is secure. If the data are correct, that is, if the visual cortex reports in comparison that the incoming data match with wind blowing along the road and not a ten-ton truck or automobile, and my aural cortex does not hear any sound indicating something coming that I cannot see, I step into the road.

Thus my conscious concern for survival determines the illumination of the cortex. It dominates the way my searchlight consciousness works. I shine the beam where it is necessary so that I will have a correct comparison between incoming information about the outside world and information associated with stimuli coming from the cortical regions themselves. Thus I compare a sight with a memory of possible visions and, somewhere in memory, assign a probability to my reckoning.

If I see a car a long way from me, there are several possibilities. It could be a car; it could be a child's toy rolling along the ground; it could be light reflecting from the heated highway. But as the car approaches and the noise of the motor increases, various possibilities are ruled out. When the car is in front of my nose, the chance that it could be a child's toy is virtually zero.

Normal waking Western consciousness has been conditioned since birth to pay attention to such things as cars, moving manufactured objects, electrical outlets, falling from high places, loud noises, and many other

[3]See Chapter 7.

danger signs. Probably through evolution we also carry a natural tendency to pay fearful attention to other people. Since we most likely survived by killing other human beings and animals, we have a built-in fear mechanism of other beings. We have been able to force our searchlight to illuminate our rational minds or our spiritual sides so that we overcome our probable tendencies to "get the other before he gets me."

Still, just a small portion of the cortex is brought into the light of awareness in ordinary consciousness experience. But when you undergo a life-threatening experience, such as having your body stretched on a rack, or taking a substance that causes you to become dizzy, or falling from a great height, and I am sure others, your searchlight mechanism no longer is able to cope with the experience. The usual technique no longer works. The searchlight widens. More of the cortex is illuminated. Cross-referencing takes place. And your body-mind begins to associate the experiences with death.

WHAT DOES THE BODY-MIND DO WHEN DEATH APPROACHES?

Now at this point there are several possibilities. If you believe that we live in one world that is objective and solid, and if you think that any experiences that deviate from this reality are hallucinations, then while you are dying or think that you are dying, you will see and hear things that are not real. In other words, you will have psychotic experiences. This is the classical medical, Newtonian mechanical explanation of what occurs. All extraordinary experiences, such as out-of-the-body, past-life recall, channels speaking through you, flying saucer visions, healings, intuitions, and whatever else you believe are not real, would fit into this possibility as merely phantoms.

However, quantum physics and relativity tell us that Newtonian mechanics is wrong. Consequently, any medical model based on it is also wrong. It is wrong at its base. It may not be wrong in its practicality. An aspirin "cures" a headache. A splint sets a broken bone. Objective reality is salvaged. But the point is, what is going on at a level below our normal objective perception? There Newton's ideas vanish in a wind of probability. That wind blows over many possibilities and treats each and every one of them as if it was all real.

A single atomic electron occupies an infinite number of positions simultaneously. It does this so that the hydrogen atom, in which it lives, survives intact. Yet there is only one electron. All of these possibilities illuminated or stirred by the wind of probability conspire to make a single objective reality. Change that mixture of possibilities, alter the wind of probability, and the reality that is shaped by it changes.

Thus, taking a second look at reality, we say that there are other possible realities alongside the one we call "reality—the only reality that is." These other realities are out there waiting to be illuminated by the searchlight of consciousness. If they do not contribute to survival, they are ignored and remain in the background of wishful thinking or fantasy.

PULLING OFF THE SCAB OF THE UNCONSCIOUS

But if survival is threatened, then watch out. The body-mind can no longer rely on the usual habitual manner of searchlight flashing. It will shine everywhere, attempting to maximize information to deal with the life-threatening situation. It will take you into other realities.

Here you will see other visions. The scab of your wounded psyche will be ripped off. After all, that scab formed to protect you from yourself. It formed so that you would not deal with or see painful memories, inept behavior, betrayals, even past-life memories that would upset you. Your ego forms out of that scab. Your personality is intact because of that wounded ego covering. You deal with the slings and sorrows of your life.

But once that life is threatened, the scab must be pulled off. You will see what a shit you have been. You will see the friends you have betrayed and cheated. You will have experiences that will lead you, too, to "clean up your act." You will also gain a new faith and insight. All who have passed near death's doorway have changed their ways.

Thus, my normal skepticism, built in by my survival programs, was short-circuited by being racked with pain or through taking ayahuasca. I could not rely on skepticism to keep me alive. I needed other data. The shamans who healed me did so because there are simultaneous or parallel realities in which my healing occurs.

I went into those realities. So I was then governed by their laws, and these laws have not been conditioned by machines. There is no mechanical thinking running the show. I temporarily gave up any mechanical thinking, as difficult as it was for me to do this, and I entered the shaman's world. As soon as I did, I was powerless to control what happened. I was simply blown by the winds of probability, the winds of ayahuasca, and my path was determined, but not by me.

To be healed there must be an alteration in consciousness. I allowed myself to be healed. My rational mechanical way of proceeding was suppressed while that happened.

THE DOGS OF SACRIFICE

As I thought back, I remembered the dogs coming into my room in the jungle. Because of their eating habits and probable infestation with fleas and parasites, those dogs would continue to scratch and fart the whole night, causing me even more discomfort. I would wake up in the fouled air and chase them from the room. One of the dogs scurried away in fear and disappeared. But the second dog did not. It left the room, but when it got as far as it would go, it lay down on the ground and assumed a submissive position to a superior dog, which is what I was to it. I pushed on her and shoved, but she wouldn't budge unless I inflicted extreme pain on her, and that I could not do.

The dog taught me about my own submission to life's experience, to the sacrifice of the ayahuasca. As I thought back to my visits with the other shamans, I remembered that sacrifice was also important in their ceremonies. The shamans are willing to put themselves through continual sacrifice in order to reach out and touch other realities. Of course, Jesus Christ did the same thing. And so did the Buddha. All the great spiritual tenets have taught us that without sacrifice there is no spiritual enlightenment.

Don Solon had been taking ayahuasca at least twice a week for forty years. You don't get used to it. In the movie *Lawrence of Arabia*, Peter O'Toole, playing Lawrence, lights a match before his army companions and slowly moves his thumb and finger up the matchstick to the burning end. He puts out the fire and does this as if he were holding a piece of grass between his finger and thumb. The dumbfounded sergeant watching him attempts to repeat the "trick." He screams in pain when his skin is scorched by the flame. "Ow, that hurts. What is the trick?" Lawrence says, "The trick is not minding that it hurts."

In other words, one goes through the sacrifice not minding that it hurts, not minding that there is pain. Don Solon cannot escape the ordeal of the ayahuasca. It involves some sacrifice for everyone who takes it.

DANCING WITH AN OPEN HEART

Soon it was eight in the evening. I looked out my hostel window. A light rain began to fall, filling the air with a mist. Ted and I went downstairs, found a restaurant, and I ate very lightly. Ted's appetite amazed me. He not only ate his own meal but ate up what I couldn't endure myself. We walked back to the hostel. I called Trisha and walked in the light rain to the house where the party was taking place. I felt both tired and exhilarated at the same time. I remained at the party for about an hour and then Trisha, I, and some other American friends, all of them doing research here, left.

We walked down a street, moving in the direction of the festive Saturday night Peruvian music we heard. Soon we arrived at a dance party and decided to go in. The party was a perfect cap for my jungle experiences. I enjoyed the music and danced for several hours. I noticed that many Peruvian men danced either by themselves or in a circle of friends. I joined in and was welcomed. Even the little boy selling gum and cigarettes joined in at my invitation, and I felt as if I was merging with old friends. I felt joyous and yet tears were in my eyes. My heart was open.

When the dance came to an end, I made my way back to the hostel and to my room. I soon was asleep. I dreamed that night that I was making love with a beautiful jungle spirit. My dream was intense and I was surprised that it was also a wet dream, a rare experience these days for me.

ESCAPE FROM IQUITOS

I remained in Iquitos for another day. I spent Sunday feeling extremely sensitive and nearly on the verge of tears. I walked around the *mercado* (market square), ate at various restaurants, and just relaxed. Although I was enjoying myself, I was extremely open and vulnerable. (Even months after the experience, I still remained so.)

On the next day, Monday, we all went to the airport. I managed to get a flight back to Lima. Ted, Jorge, and Don Solon were due to fly to Tarapoto around the same time. But their flight was canceled again. Later, after I had returned to the United States, I spoke with Ted. He told me that all flights between Iquitos and Tarapoto had been canceled for the rest of the week, so Ted never returned to Tarapoto after all.

I was finished with ayahuasca. I had learned what I needed to know.

THE CABINET OF DR. CABIESES

I arrived in Lima quite late, since my plane didn't leave Iquitos until nearly ten that evening. I took a *collectivo* (a van that carries several passengers) from the airport to my previous hostel, Sandy's Hostel. The hostel keeper was quite surprised to see me. She was expecting me on Wednesday. My room was ready, however, and soon I crawled into my now familiar bed and was asleep.

The next morning, I telephoned Nora de Izcue, the film-maker I told you about, living in Lima. I also wanted to speak with Dr. Fernando Cabieses Molina, who ran a neurological clinic at the Museo de la Nacion (national museum) in Lima. Dr. Cabieses was the main organizer of the Traditional Peruvian Medicine Conference held last year in Lima.

Jorge had told me about Nora, the director of the film *The Winds of Ayahuasca*. Later that day I met Nora and her companion Sandra Weise, also a film-maker, at a small Italian restaurant. We talked about my experiences, meeting Jorge and Don Solon, and my days in Peru. Nora told me that she had recently gotten a copy of her film subtitled in English and that she would be happy to show me the film the next week.

A week later, after my trip to Cuzco, I was back in Lima. (You'll remember that I went to Cuzco to experience the Inti Rahmi ceremony that I had dreamed about before I arrived in Peru.) I wondered about my last two contacts. I called Nora and I called Dr. Cabieses. Cabieses was extremely busy but agreed to see me the next morning at 8:00 A.M. I agreed to meet him at his offices.

I arose early that morning. For some reason I was quite eager to meet with Dr. Cabieses. I took a taxi to the old Museo de la Nacion on Avenida Javier Prado.

There were armed soldiers and police surrounding the building. I went up to a guard, announced myself, and the guard directed me to the elevator and said, "*Piso octavo*" (eighth floor). I rode up and got off as instructed.

I found no one there. I managed to find an open office and I asked, "*¿Donde es la cabaña de Dr. Cabieses?*" (Where is the office of Dr. Cabieses?) I probably used the wrong word for office, but it reminded me of *The Cabinet of Dr. Caligari*, an old horror film about a mad neurosurgeon.

Around 8:15 A.M. a guard came in and opened Dr. Cabieses's "cabinet." I went in, sat down, and was offered coffee by a receptionist. Soon a secretary came in and led me to the doctor's office.

His office had a grand view from the eighth-floor window. I found him sitting alongside his conference table. He was just placing the phone down. He welcomed me in perfect English and told me that he had only a half hour to talk with me. I asked him how he happened to have an interest in ayahuasca.

He replied, "I have done a lot of work regarding the archeology of the use of hallucinogens." He then told me about the many substances he had investigated. I didn't recognize any of them. He said, "One of the substances was *wilca*. This substance is made from the pod of a tree. It produces a lot of mucus and a lot of tears. This is why many of the shamans depicted in the ceramics have long tears flowing. They used this wilca to communicate with the gods. I have researched this and I have written papers about it."

I finally decided to ask him directly. "What do you think happens in the brain when you take ayahuasca?"

THE MODEL OF DR. CABIESES

"I only have a model to explain this. It is not original with me. I think that all your mind is like this room. It has files on the walls. And everything in the room is dark. Your consciousness is just one beam of light. Where you illuminate a file, you have a memory. To remember your mother's face you have to look that way. But then something you seek is not there easily. It is there, yet you cannot find it."

He continued, "For example, you want to remember a song that was played in such and such a movie. You can't remember it, but it is there. You know that, because when you hear the song you recognize it. Or suddenly you hear the song when you are taking a shower. But when you look, you cannot find it. The reason is that the beam doesn't know where it is.

"But when you take ayahuasca, everything illuminates at the same time. Then you know where everything is. You just look around and find things easier. Things come to you without you even wanting them."

"Could we say that the searchlight beam becomes more diffuse?"

"No. I would say that the file illuminates itself. It is constantly working. That part of the brain communicates with this part without you knowing it. That is how you solve problems without knowing. That other part flashes once in a while. Then if it reaches you, the little song comes. It flashes. With ayahuasca, everything flashes. Every file is lit up. You don't need the beam anymore. It is not the beam that becomes better, it is the whole brain that ignites."

"So is the beam in your model associated with some will or intention?"

"These files work constantly. You don't know what's going on between them. There are communications and associations constantly going on. You need the beam to know about them. If you put out the beam, you go to sleep. And everything continues working. Then you flash up your beam and then look here or look there, constantly looking at something."

"Is the beam associated with intention or will?" I repeated.

"Yes, definitely."

"Then what is the difference between being intoxicated with ayahuasca and normal waking consciousness?"

"With ayahuasca, everything illuminates itself. You don't need any beam."

"Without the ayahuasca, you have the beam?"

"Yes. This beam of light makes you one individual. You have just one beam. If you had several beams you would become schizophrenic. If everything lights up, you are crazy until you learn to work with that. You become overwhelmed with light when you first take ayahuasca. The shamans take it for four or five years. They train themselves to work with the lights on. Then, once they understand this, once they know how to work with every light on, they can find many things. They can divine."

He paused. I was thinking about the future and how ayahuasca enabled one to see into it, but I said nothing. Again, Dr. Cabieses surprised me with his apparent psychic ability.

He said, as he pointed out the window, "What is learning about the future? It means that there are clouds here. I know that when there are clouds here, and there is wind coming from that direction, we will have rain in three hours. This is divining the future. But this is logical. It is a logical deduction. It is cause and effect."

"So you feel that the power of the Ayahuasquero is a power of logical reasoning with an access to a greater amount of information that is normally not available to him?"

"Yes, and it becomes available simultaneously. You get availability to all your files. If you learn to do that, and it takes nearly four years, then you are a better man. You can learn to do that through drugs, through ritual, through meditation, through any kind of physical denial like hunger and cold. Fasting does it too. It causes a lot of biochemical situations that put your brain into different states."

WHAT IS YOUR CONSCIOUSNESS, AND WHERE IS GOD?

He took a breath and went on. "All of this is altered states of consciousness. What is consciousness? It is the beam. The beam is it. You have only one consciousness, which is the beam. If you light up all your files, what is the beam needed for? You forget about the beam. You just sit down and see. If you learn how to orientate yourself into this tremendous file of your memory, then it's okay.

"You cannot hallucinate on something that you never saw. You hallucinate on things that are in your subconscious or unconscious mind. The shaman will hallucinate with a jaguar. You will hallucinate with a Jaguar car."

We both laughed at the joke.

"Okay," he continued laughing, "because it is in your consciousness, you are prepared for that."

I told him, "Well I had some experiences with ayahuasca that I wasn't prepared for. I had the experience of flying as an eagle looking down on the ground."

"But you have seen many movies of that. You have flown, not as an eagle, but you have flown. You probably connected things like flying, eagle, the symbolism of being an eagle, the actual feeling of flying, the pictures you get from being on a plane. You get all of them together and then, boom, it comes to you."

The explanation certainly made sense to me. But still the process must be incredible.

He continued, "There is an acute discomfort between religious people and people like you and me who just like to see the objective things. When we say that God is inside of you, that God is a product of your brain, that you created God, that man created God to his image instead of God creating man to His image, there is always a conflict. People say that I am an agnostic and I am a materialist and God is really there.

"I always answer, suppose that you are God. And you create a little creature and you want to communicate with that creature, are you going to appear to him all big and bearded and things like that? You would get into his brain. You get into his brain, and then if there is God, God will be there. You may feel that you created God, or you may feel that God created you; it doesn't matter. This is semantics.

"The actual thing is there is a communication between you and something that you can see is supernatural. The supernatural is in your files."

I explained the connection between quantum physics and consciousness called the observer effect. I told him that consciousness must have something to do with that. The act of consciousness is perception. In a normal waking consciousness, that choice of perception whether to observe momentum or position alters the complementary state. If I measure location, all information about momentum is no longer accessible to me. If I measure momentum, all position files are dumped. I can't know or illuminate both sets of files at the same time. They can't both be objective simultaneously. One or the other is objective, but at the expense of the other.

However, in a shamanic state of consciousness, it appeared that not only two files could be seen simultaneously, but the whole file cabinet. This leads to a role of the searchlight that is different in the two states of consciousness.

I then asked him, "How does one objectively deal with this?"

"I'm thinking that the problem is that you are examining that position-momentum dilemma with your consciousness. But this is another dilemma. Actually it may be that you are inventing the whole thing. There may be no such a thing. So it is impossible to grasp."

Finale: It's Just a Movie, Fred

Father, speak to us of the winds of Ayahuasca, tell us what you see. Let the winds of Ayahuasca come to the earth so that we may drink from it, then we will go there . . . up there. . . . Why do we have to remain here?

(Fragment from a ritual taken from the Linguistics Group called Orejon from the Peruvian Amazon)

Dr. Cabieses laughed and told me that he had no more time to spend with me. He thanked me for coming and I left the office. I returned to my hostel, went out for lunch at a *chifa* (Chinese restaurant), and walked back to the hostel to wait for Sandra Weise. She was coming to take me to the movies. And you'll recall (Chapter 2) that it was Sandra and Daniel Pacheco who picked me up and took me to the movie *The Winds of Ayahuasca* that started my story.

And now in the cab making my way to Jorge Chavez Airport in Lima, I realized that the story was over. Or so I thought. There were still some pieces of the puzzle I had to assemble. The wind had stopped blowing. I asked the cabby about sudden winds in Lima at this time of year. He was as surprised as I was. As the cab continued its journey, I went over each hypothesis. I realized that the order in which I originally made my hypotheses did not match the order of my conclusion. It didn't surprise me, considering what I had just experienced. I began my thoughts with the seventh hypothesis:

Hypothesis 7. All shamans work with a sense of a higher power.

Shamans seemed to be born, not made. Shamans were sensitive to the earth and enhanced that sensitivity by using sacred plants in their healing rituals. In Chapter 5, Paul Devereux confirmed that geomancy indicated the sensitivity of people to place. You'll recall my speculation that shamans were probably those individuals who were most sensitive to the anomalous energies in their environment.

Furthermore, it was apparent to me that the big difference between modern religion and shamanhood was where God was put. Shamans all sense their higher powers in terms of earth spirits as well as heavenly spirits. Remember that Jamie Sams and Doug White evoked the spirits of the four directions, the sky above, the earth below, and with Jamie, the spirit within. But being a physicist, what interested me more was my discovery that shamans' powers may have come from anomalous radiation found in sacred sites around the world and from the presence of mind-altering plants that seemed to be connected with the locations of these sacred sites.

The Brighton shamans made this clear to me. They pointed out that the various devices used by shamans had limited power and worked best in the countries of origin. Richard Dufton said, "We are the human organisms grown on a particular patch of land. We eat that land. Our mothers ate it. We are that which our mothers ate and that which our forefathers were." Shamans remain sensitive to their own land vibration wherever they happen to be.

I was convinced that the shaman's power was found in vibrational energy and that the vibration was connected to the vibrational patterns of the shaman's own country. This led to:

Hypothesis 1. All shamans see the universe as made from vibrations.

The underlying reality that shamans deal with is vibrational. They deal with a world of vibrations, cycles, and circles, not straight lines. The key to understanding their world was to realize that it was a vibrational world—a world of sacred songs, chants, and rhythmic drumming—all designed to alter the consciousness of the patient, enabling that patient to heal by taking the patient's mind off his or her normal waking consciousness and into a world of myth.

The major point in this was recognizing how the healing occurred in terms of a simple physics model of vibration. I believe that shamans are capable of producing within their own bodies and/or nervous systems certain vibrations that are capable of absorbing illness from a patient.

Jorge had made this very clear to me. The shaman *must* take on the illness of the patient in order to heal him. But how did this occur? I believe that illness is found in the vibrational patterns of the body. Thus an ill person's vibrational pattern has contained within it, literally, sick vibrations. These could be spurious frequencies or amplitudes. If they involved the nervous system they could be vibrational patterns superimposed on nerve impulses.

To absorb these spurious vibrations the shaman must be able to pick up the ill person's vibration. This was nothing but a resonant interaction. The shaman and the patient were capable of exchanging energy, provided they both were able to vibrate at the same frequency. Shamans knew how to induce this vibration within themselves. But an ill patient did not. Since the shaman's vibration is strong and determined, he must somehow adjust the ill person's vibration in order to resonate with his own. Then he must take the ill person's negative energy from him.

But in order for that to take place, the patient had to undergo an initiation that enabled him to produce a similar vibration in his own body. Shamans then healed by absorbing the illness through the resonant interaction with the patient. The shaman had the healing frequency. The patient, by undergoing a ritual, learned to get in phase and frequency with the shaman and thus transfer his illness to the shaman.

Since the transfer of energy required a resonance between the patient and the shaman, shamans were vulnerable to other people. Normally shamans conducting ceremonies absorb illness from the patient and then dissipate it through urination, bowel movements, spitting, throwing up, and the like. However, if the patient sends a vibration that the shaman is not expecting, the shaman, particularly if he is vulnerable to some illness himself, can become very ill. This explained the clouded eye of Don Solon. Apparently he had vision problems that another shaman knew about. Jacques Mabit also had told me that by changing his thoughts during a ceremony he actually caused a shaman conducting a ceremony to become ill.

Jorge had told me that there were three different types of magical healers: curanderos, shamans, and brujos. Curanderos worked with white magic; they healed using herbs, chants, and other techniques. Brujos dealt exclusively with black magic; they cast spells on others. Don Solon was injured by a brujo. Shamans worked in both realms; they healed and they could cause illness, even death.

Thus the vibrational energy transfer of illness could go in either direction. It was a two-way street connecting the shaman and a patient.

In the best situation, the resonant connection between the shaman and the patient that allowed the energy transfer was felt by the patient as loving energy. This led to:

Hypothesis 8. Shamans use love and sexual energy as healing energy.

In Chapter 6 I explained that love and hate were forms of vibrational energy. Photons, particles of light, acted as love—a desire to join in the same state. Electrons, particles of matter, acted as hate—a desire to separate. Sexual energy was the dance between the two. We are a composite of an electronic and a photonic body. To heal, a shaman, acting as a curandero, would induce a resonant condition in the light or photonic body. Particles of light tend to act inclusively and enter into the same resonant state. I now concluded that the light body inside of each of us was where that resonance existed.

Making an analogy, if we separate the electronic body from the photonic body, we actually have three bodies, not two. There are the electronic body, the photonic body, and the body in between the two. We might call this the quantum body of consciousness. It was able to work as the healing shaman inside of each of us. The electronic body dealt with survival and was capable of destruction. Separation and hate were its domains. This was the domain of the brujo. When the brujo cast a spell, he affected the electronic body only. Thus a brujo's action was strongest on those who felt most unloved and least sexual.

The curandero acted to heal, and thus acted on the body of light—the photonic body. The more a person accepted himself or herself as a body of light, the more loving the person was, the greater was the curandero's action.

The shaman acted on the quantum body. He orchestrated the dance between the photons and the electrons. This would be felt as sexual energy in the body. In brief, I saw love as the curandero, hate as the brujo, and sex as the shaman.

So shamans were connectors, and love was a connection energy. Being "in love" enabled one to feel connected to the whole universe. In a mythic sense, shamans were reminders that we all are connected. We are in a very real sense one consciousness. Shamans were aware of their connection with the whole universe. But we living in the Western world may not always feel this connection. Sometimes we had to be tricked into believing that it was true. For this to happen there was also a confidence game going on. This led to:

> Hypothesis 4. Shamans use any device to alter a patient's belief about reality.

To be healed by a shaman was actually simple if the patient believed in the shaman. That belief would enable the patient to reach the vibrational level of the shaman, enabling the illness to be transferred. But not all patients were able to believe in the shaman's powers. Those patients had to surrender their belief systems. This was the reason shamans used tricks, when they felt they were necessary. You'll recall Holger Kalweit's story about quaffing down hot butter lamps, the story of Doug White and the flashing lights during the Sioux ceremony, and my experience firewalking. These experiences seemed to violate the laws of physics.

Probably there was a rational explanation for all of these. I tried to explain the firewalking. I was not able to explain the flashing lights in the Sioux ceremony. Holger did explain the hot butter lamp quaffing.

Yet, there was more to these experiences even if they are scientifically explainable. By believing in them, the person witnessing them undergoes a reality shift, even if later on he or she recalls the experience as a sham.

Shamans were comfortable with a reality that violated the laws of physics. Most Westerners weren't. Again, we had to be convinced to change what we believed. But shamans were already convinced. This led to:

> Hypothesis 2. Shamans see the world in terms of myths and visions that at first seemed contrary to the laws of physics.

The shamanic world usually is a "third world," an often primitive and impoverished one. It is a world where the objects of Western technology are either absent or of lesser significance. We come from the world of technology. So for us an alteration in belief is necessary. The key insight comes from extending our belief system beyond the current four walls, a ceiling and a floor, and the grandfather clock ticking in the corner. We must begin to see that there is a whole world, a real world that exists beyond space and time as we presently know it.

This world includes the richest experiences that human beings can have. This is the mythic world, the world, perhaps, of archetypes or ideals as seen by Plato. Some researchers now call this the "imaginal realm." Whatever we call it, it certainly is a world that I, as a physicist, had to consider in my own cosmology, although perhaps in a limited way.

For example, we have never seen an electron, but we believe that it exists. We have no idea what it looks like. Its behavior is bizarre, if we try to describe it in terms of familiar objects. Our picture of it is in our minds, only, and yet, with that picture, we build up a vision of how the whole universe is constructed. In this sense, physicists tap into the imaginal realm in making models of the invisible world. My research in the shamanic world suggests that we can go much farther into the imaginal realm.

Healing is a mysterious process. My model based on vibrational energy transfer is only a model. It could be wrong. No one really knows how the body heals. How can a cell on the surface of my thumb heal a cut there? We don't know. But it can. The shamans see the healing in terms of a spiritual presence. That is their archetype. The quantum physicist hopes to see the healing process in terms of quantum waves of probability. Perhaps these are just complementary ways of seeing the same thing. I think that they are one and the same.

If we manage to see that they are the same, we may find an adequate model for healing based on archetypes or mythic elements, in much the same way that physicists found an adequate model for matter in terms of atoms. No archetype will perfectly fit a situation. In much the same way, no model of an atom fits any classical perception of a solid object. Atoms are not things, as Heisenberg reminded us a long time ago. Neither are archetypes. But we believe in atoms and think of them as things. We need to believe in archetypes in the same manner.

Thus I saw that the shaman entered a mythic world to heal a patient in much the same manner that a physicist entered a laboratory. Each did so in order to enter a state of consciousness consistent with their tasks. This led to:

> Hypothesis 3. Shamans perceive reality in a state of altered consciousness.

The mythic and the temporal worlds were overlapped for the shaman when he was conducting a healing ceremony. Thus shamans would see spirits of departed relatives, plant spirits, visions of past and future in a kind of hypnagogic state. They were able to maintain that state of consciousness for quite a long time. Most were able to do this with hallucinogenic plants as their "guides"; however, not all needed them.

Yet, it seemed to me that the plants were vital for really serious healing in our Western world. Although I may be in deep trouble suggesting this, I believe that the Western world must begin to take a more tolerant look at sacred, vision-producing substances, particularly when these substances are taken under the guidance of a shaman—a person of knowledge about the plant world.

I couldn't even conceive of taking ayahuasca as a recreational drug. It would be far too dangerous to do so. But I believe that this substance could be used by the medical profession, with the participation of Ayahuasqueros, to heal a number of serious mental/body illnesses. My thoughts go out to the number of drug addicts in our society. I believe that a program of monitored ayahuasca journeys for drug addicts would enable drug addiction to be cured. I also think that ayahuasca would be useful for curing depression. I am thinking in particular of the many recovering Viet Nam War veterans, who I understand, after some recent research gathered by spending time with veterans from VA hospitals, are in serious mental straits. Let me explain why I think ayahuasca could serve this purpose by recalling some of my own observations of myself when I had taken it. I had been talking with Dr. Cabieses in Lima.

He told me that on ayahuasca your whole brain becomes illuminated, and nothing in there has come from anywhere but your own mind. All the experiences, past and present, become illuminated at once. This makes you crazy. But this gives you insight.

I then remembered the several photographs I took in Peru. I like to take panorama shots, with one picture overlapping onto the next when they are laid out on a table. The whole vista is there, but the choice of what can be seen and remembered is mine. Taking a series of shots of the skyline over Machu Picchu, one must choose what will be recorded. Normal consciousness is like the photographer who shoots selected scenes to be remembered. With normal consciousness a single scene at a time is illuminated.

With ayahuasca, the whole sky of the unconscious is illuminated at once, as if the photographer had a super omni-camera able to record the total global sky picture at once. Not just one scene following one scene after another, but all at once, now and forever.

Think of the whole field of consciousness as the hemispherical sky. I saw normal conscious perception as if it were through an omni-lens seeing everything, but growing cloudy as time passes and hurts are recorded. The cloudiness is caused by a huge scab covering that omni-lens. The scab covers and occludes everything from my view except a small bit of light.

Under ayahuasca, the scab is suddenly dissolved and I can see everything at once. The wounded psyche is exposed to the winds of ayahuasca, and for brief moments my whole unconscious becomes alit. But then the scab begins to form again, doing what all scabs do, protecting and preventing the outside world from reinfecting the wounds we all feel from everyday encounters with ourselves and others.

I still remember what the exposure to the wind has taught me: I no longer needed to protect myself from that wound by reliving the pain of that past experience over and over again as I move through my life into new situations. This time I move with opened eyes and without the blinders of scab protection. This time I don't act out the role in the same way that I did. I don't re-create my life in the same old pattern. I am born again in my unconscious.

But the removal of the scab is painful. Seeing myself as I am is painful. We all live in our illusions of our own mastery of life. Even if that mastery is failure, we are masters at finding excuses for those failures. If we are rational to a fault, we will argue ourselves into the pit of our own despair over and over again, forgetting what the winds have brought us: pain and vision.

And now it is in the remembrance of those feelings that I see how the vision vine helped me to see my own psyche in a new light. I believe that anyone suffering, such as drug addicts and recovering warriors, would experience a similar sense of their own destiny and recovery by taking a shamanic journey. They would be able to reconnect the events of their lives in a new and more meaningful manner. If this happened, the underlying cause of the illness would surface, and by becoming aware of it, a patient's illness would vanish.

In brief, the shamanic state of consciousness, as provided by ayahuasca or any other means to induce a shamanic awareness, enables the person to see himself in a mythic light. This vision provides a sense of compassion, a connection with all life, and a new reason for existing. But to gain this understanding we needed to make physical life more meaningful. This led to:

> Hypothesis 5. Shamans choose what is physically meaningful and see all events as universally connected.

Let me get to the universal connection first. The key physics concept here is the hologram. In modern terms, shamans see the universe as a gigantic hologram. You'll recall that a three-dimensional holographic image of an object looks so real that, for all intents and purposes, at least visually speaking, *it is real*. However, and this is most important, that visual information, frozen in the hologram by captured interfering light vibrations, is strongly dependent on the observer. By changing the direction or the frequency of the reference beam shining through it, the information perceived is altered. By moving the reference beam around the hologram, different flashes of information in the form of pictures are perceived. And by changing the frequency, the images change shape and size.

You'll also recall physicist David Bohm's vision of the universe. He also saw it, like the shamans, as a hologram that he called implicate order. When an observation of an image in the universal hologram occurs, Bohm says that an explicate order has been experienced.

In my view the explicate order is controlled to some extent by the choices that the observer makes. However, the original hologram, the implicate order, contains far more than is obvious to the observer; every act of observation, while it brings forth a view of some objective quality, at the same time renders other objective qualities invisible.

And now let me get to the first part of the hypothesis: meaning. I discussed some of this in Chapter 7. The key here was to realize that shamans manipulated matter and energy simply by altering the ways they observed it. They were able to change the significance of any two events, i.e., the relationship of one event with another.

Now according to quantum physics, the great world of events that we witness in our everyday lives can be ordered in pairs. Event pairs are required to have meaning in the world. A single event can take on meaning only when compared with another single event. The more comparisons that can be made between a single event and the other events, the stronger the meaning associated with that event.

For example, an initial event might be the impact of a hand on a drumskin, signaling a beat. Another initial event could be the flash of light as the sun sinks below the horizon. The final event might be the vibration of the eardrum of a shaman.

Of course, there are many events taking place in our everyday lives. What we call *meaningful* is the connection of these events. The impact of a hand on a drum followed by a vibration of an eardrum is probably more meaningful than the connection between a flash of light of the setting sun and the vibration of the eardrum.

If we take *meaningful* to mean the pair of events having the greatest probability of connecting, then the hand on the drum followed by the vibration of the eardrum is more meaningful than the flash of the setting of the sun followed by the vibration of the eardrum.

However, shamans could change all that. They were able to make the sun-flash connect with the vibration of an eardrum. Not only could they do this in themselves, they could alter a patient's consciousness so that he or she experienced the connection also. In modern psychology, one example of this is called the crossover of senses phenomenon. I explained in Chapter 5 how shamans can sing certain songs that provide a crossover into the visual field. The songs induce visions of geometric design. I also experienced this crossover in several ceremonies and described this briefly in Chapter 7. Jacques Mabit described his crossovers in Chapter 9.

But shamans could induce more than just crossovers of senses. How did shamans alter the connection between events? Before I explain that, consider how we make the connection between events in normal waking consciousness. This can be made clear from my embellished vision of the Anglo-Saxon shamans.

We can think of any event-pairs as two ends of a length of a strand of the infinite spider web—the "web of wyrd"—of the Anglo-Saxon mythology. We can imagine that our consciousness moves through these events like a spider making its way along the web. Just as our consciousness chooses which events to pay attention to and which events to ignore, determining which events are past and which are present, the spider moves along the web of consciousness usually choosing to crawl along the strongest strands. The strongest strands are those that connect events with the greatest probabilities. The strongest strands are to be trusted. They have been woven and rewoven because the spider's survival depends on those strands holding up. Our minds are like that. The strands in our minds are the least-action paths of habit that we have created.

In Chapter 7 I explained that least-action paths, like the strong strands in a spider's web, were the paths of habit. By changing one's way of seeing, it was possible to change those paths. Candace Lienhart had made it clear that these least-action paths became or constituted part of the unconscious mind and particularly that part of the unconscious that dealt with survival.

This technique of altering least-action paths between events and thus changing the meaning of one event with respect to another is accomplished through the shaman's imaginal senses. I told you about them in Chapter 7.

It was the shaman's power to change those pathways that made me excited. In Chapter 11, I explained how they did that by altering the connection between two events simply by observing a midpoint. Normally between any two events, as I explained, all possible paths are in existence. These constitute what are called parallel worlds in physics. When no observation is made between the pair of events, the least-action path is remembered as the "real" pathway.

But by making an observation "in between," a new pathway is made and the connection between the two events is severed, replaced by the third mid-event. The mid-event is now least-action-wise connected to the previous events, and that becomes the remembered sequence of events.

I remembered my purification experience with Jamie Sams in New Mexico. I described this in Chapter 4. When Jamie noticed that the other people were not coming to the ceremony, she immediately connected this to the presence of the three women with myself. She saw that this was destined. The others were bound to be lost.

Normally I would have connected the event of their not coming with the fact that they got lost. Jamie brought in the connection that they got lost in order to have exactly four people, herself, two women, and one man needing feminine spiritual healing, present. She saw the balance between the three women and myself as a mythic necessity, probably dealing with the spirits of the four directions.

Often unexpected events leave us at a loss for explanation. Usually we are dealing with two or more events. But by bringing in a mythic event the explanation becomes clear. Let me give you another example of how this works. Suppose the first event is my mother leaving me alone when I was a child. Suppose the second event is me leaving my child alone. A logical connection between the events is that since my mom left me and I survived and am perhaps even stronger because of this, I can leave my child. But suppose a mythic third event is put into my consciousness. Let that event be the conflict between Eros and Psyche. Eros left Psyche and she lost her love for him. She regained it only after a long search and much suffering.

Now connect up the dots. Mom leaving me connects to Eros leaving Psyche. I then realize that I suffered and no longer felt the love for my mom. Now connect up Eros leaving Psyche with me leaving my child. I realize that I will suffer again because my child will stop feeling love for me.

Now take into account the mythic connection between you and your child and between you and your mother. Perhaps the complete myth of Eros and Psyche would even add more to your connection with your child and your mother. There are, of course, other examples one could think of, but I hope you get the point. By introducing a mythic element into consciousness, reality is changed or more of it is revealed to you.

Now those mythic elements were found in the shaman's belief in the existence of parallel worlds that included the past, the future, and the present in all of them. And it included bleed-throughs from the mythic world into the physical world. This leads to:

Hypothesis 6. Shamans enter into parallel worlds.

First let me point out that in quantum physics the parallel worlds we speak about are physical. According to this model, we are always in the illusion that we exist in just one world, even one universe, when in fact we exist in an infinite number of them all of the time. These worlds are constantly interacting, coming together and flying apart in dimensions we cannot experience, every time anyone makes a single observation. It is just the action of shifting consciousness that causes one to decide or to think that one is in one universe or another.

The universes or worlds all overlap and form a hologram. Shifting from one world to another is just changing a viewpoint in the hologram. In Chapter 8, Jorge told me that he was able to journey between parallel worlds. In one world he was at home, in the other he was with a sick child. I saw these as parallel physical worlds. In the world where Jorge was at home, he appeared as a ghostlike image to those who were in the home of the child. Jorge made this journey after very special preparations. Kote Lotah was able to put his consciousness into a raven and observe his wife driving a car. Even I have had dreams of waking up in an animal's body, particularly that of a bird.

And, of course, my lucid dreams were, I now believe, visits to other worlds in the same sense that Kote and Jorge were able to make such journeys. While I could not completely explain this to myself or to anyone else, I think that the parallel worlds concept leads one closer to this possibility than any so-called objective reality theory of physical existence.

It was the time-traveling element that really amazed me. Now I did explain in Chapter 9 that every shaman I spoke with and spent any time with confirmed his ability to travel in time. They all claimed to glimpse events in the distant past and the future. I now realize that since shamans can do this—travel outside the limits of space or time—then clearly there must be something outside of space and time. There must be some kind of reality that exists "out-there" and "else-when."

In Chapter 9 I explained how, through the action of the quantum wave, it was possible for a future event to connect, by traveling backward through time, with a present or past event. I also tried to explain that perhaps it was visions of the overlaps of different time periods that constituted a myth. Myths were just messages from the past and/or future all superimposed.

But now I wasn't so sure that this was correct. I suspected that there were really two parallel connected realities. All that I wrote about the parallel physical worlds can be lumped into one unifying concept: chronos, the world of experienced time. Everything else I called mythos.

These two worlds were much more strongly connected than I ever thought before. The worlds of chronos and mythos were as connected as the two poles of a magnet. Attempting to rid oneself of the mythic world was like cutting a magnet in two to get rid of a north pole. As soon as you broke the magnet, a new north pole would appear at the break. Mythos and chronos were complementary worlds.

Bohr's principle of complementarity echoed in my mind. When one is living one's myth, there is a flow, predictability, certainty, and knowledge of the future. But paradoxically, there is little sense or awareness of the material universe. You know where you are going, but you don't care.

On the other hand, when one is living fully in space-time-matter, there is nothing but insecurity, unpredictability, scarcity, and suffering. One cares very much about what is going to happen.

The ultimate experience of full mythos realization took place at death. This led to my last hypothesis:

> Hypothesis 9. Shamans enter the death world to alter their perceptions in this world.

I then realized that the shamans had been teaching me a way to shift my perception so that I became aware of other realities. Although I hadn't fully realized it when it happened to me, the shamans had brought me to the doorway of death. No one will disagree that death alters one's consciousness. And since our main focus of consciousness is on survival, I would expect, if my theory was correct, that when one is dying the focus of consciousness will no longer remain on the body survival reality.

It won't because the body is telling it that that focus is no longer working. It is like looking for a set of lost keys under the lamppost simply because it is lit up. The keys are not there; they have been kicked into the darkness, but we like to look where the light remains. The body and the mind probably play this way. And it works so long as the body-mind is convinced that this strategy is successful for survival.

I had been through a number of shamanic experiences. The most powerful involved my taking the vision-producing substance ayahuasca into my system. And I had become ill and actually felt that I was entering the domain of death.

Now, I believe that drug intoxication convinces the body that it is dying. There is probably some evidence that near-death experiences can be related to drug intoxication. Under such an influence, the normal survival instinct is subjugated, and the focus of consciousness is willing to grab for straws on the ocean, since the lifeboat is sinking. Or in another metaphor, the ice of the ego surface has melted under the heat of shamanic awareness. In other words, the mind will engage less probable realities because the mainstream reality is no longer leading to the survival of the body.

In the ayahuasca experience, the body seems to have a near-death experience. The body-mind—the mind that is totally occupied with the circulation of the life force, the blood, the nerve impulses, and so on—begins to get the message that something is wrong. When the body-mind is threatened and thinks it is going to die, it will engage other realities, seeking for its own continuance.

When that happens, these other realities that are present appear to us as a vibrational resonance of another reality with our own reality. For a moment our attention has shifted. It is like using a camera to focus on the background for a long time and then suddenly pulling the focus in to see the foreground. These resonances occur for a few brief moments for an initiate. For the shaman, they last a lifetime. As the body processes ayahuasca and the effects of the substance dwindle, the body gradually returns to normal reality. The focus changes from foreground to background again. When the body really dies, the other realities appear with greater frequency, and when death is fully present, these other realities become totally real.

That was it. In a nutshell, shamanic physics consisted of all the experiences in consciousness that result from seeing the universe as a gigantic hologram or spider's web. By extending one's belief system, it was possible to reconnect with the whole universe—to become one with everything and be healed. Shamans did this through the observer effect; they altered reality by altering the way in which it was observed. Invariably they brought the mythic and death worlds into this one to do this.

That was my model as simply stated as I could. As the cab continued, I thought about Dr. Cabieses's "searchlight and file cabinet" model of consciousness again. He did tell me that the model he described was not his own. But he failed to tell me whose model it was. I felt that it was a good model to explain the action of consciousness, and I liked the idea of thinking of the brain as a room with files neatly stacked along the walls. The idea that consciousness was a searchlight beam, and that the recall of a difficult memory was due to the item being hidden in a corner of the room, appealed to me.

However, I did not like his materialist bent. Perhaps this bent comes from performing too many neurosurgical procedures. Perhaps one gets a little jaded. But then again, he was clever and I liked him very much. His materialistic determinism, however, echoed in my head. Could the universe be determined in some way after all? Quantum physics forbade it!

I began to think about Jacques Mabit and his theory about there being two times: a chronological and a mythic time. We live our lives in Western society as if there was only one time—the "real" chronological time. But we live our lives in such time vacuously. We live according to the clock and we become machines.

But the real life, the real juice of life, is not to be found in the clocks. It is found in our myths and our fantasies. Somehow the film had interwoven the two times for me. My myth and my reality were at that moment one, pasted on top of each other like the overlap and superposition of two reels of film.

I saw that the film had been scripted. Once it was on celluloid, that was it. One frame followed another, and no one could change a single word of it or a single scene. The film, carrying through a story, was completely determined. Yet the associations of each frame with each other were entirely in the minds of the viewers. The viewer made sense of the film. Life is how we see it.

The cab pulled up to the airport entrance. It was still very early in the morning. My flight wasn't to depart for another three hours. I made my way through customs, airport tax collectors, and into the departure lounge. I sat down and waited.

And now, when I think back over my whole shamanic journey from the very beginning, as I wait here, in the early morning, waiting at the Jorge Chavez Airport for my flight back to the United States, I finally see clearly what I was missing.

I had tried to reconstruct everything from the world of chronos. But I had sensed my own mythic being. I had felt that the story I had been living had been written by some spirit, some ghost-writer in the sky, and I was just an actor playing a part, carried along by the winds of ayahuasca.

When I think about the difficulties of just getting into and out of the jungle, how all of my flight plans to change the original itinerary were altered, how the jungle just swept up to meet me in a flood of rain, canceling out my flight alterations, making it impossible to leave Tarapoto for Iquitos and, then later, fly from Iquitos back to Tarapoto, where I was planning, very reluctantly, to take another ayahuasca journey, I see that I couldn't change a single word of the script. I had to submit to the wind and the rain of the jungle spirits.

When I think back to my meetings with the young Peruvian woman, Nonoy, and the song of the shaman, "noh noh noh noh . . . ," was the song a warning telling me "no" to Nonoy? I see that she too was an actor playing the part of the spirit of the ayahuasca. When I remember her coming to collect me at the casa de Jorge, I still see her small body and her tiny pointed breasts. Was it any wonder that I was attracted to her? Just picturing her and remembering my final departing at the airport at Tarapoto when she failed to meet me, was there a misunderstanding? Did I do something wrong? Or was it written that way? She was beautiful and she was so attractive. Was she sent to attract me? What role did she play? Maybe I played my role the way it was intended to be played.

I didn't want to be involved with her, but somehow I felt compelled to approach her as a lover and be refused. Was I playing the role of a seeker, attempting to seduce the spirit of ayahuasca? Or was this innocent relationship just a fantasy induced by the fact that we lay together under the jungle canopy swooning to the effects of the ayahuasca and the songs of the shaman that night in the balmy air of Tarapoto?

And then that day I left Tarapoto, those drunken Peruvians behind me, pushing me, in the line getting on the plane, were they rude people or spirits reminding me again, no, no, Nonoy? "She is our spirit, not yours to be captured." They called me "*tio*," uncle, a term of endearment or was it an insult, a reminder that I was old, too old to have the young spirit of ayahuasca as my lover? I turned around to them, "*Me no tio*." I'm not an uncle, I'm young and fit. I'm not an old man. But I am becoming old whether I believe it or not.

Inside me I believe I was just an actor playing some heroic part in a jungle movie. The fact that Nonoy herself was an actor in a Garcia Lorca play in Tarapoto—she asked me to watch her perform at a rehearsal—was a peek behind the curtain of the play that I was in. She was playing a lead role in the drama. Her leading man was introduced to me. He smiled and wondered, was I her grandfather? Realizing that somehow my spirit was seen differently than I saw it enhanced the vision that it was just a movie and I was another actor. I was playing a part in their lives that I fully failed to realize. Was I a father image? My illusion was just a part of this bigger illusion.

Did Nonoy mean she was an actor in my play? Why else was she there at my first ayahuasca ceremony? Why did she take her place next to me? Did she know that we would share the experience together? And later after the ceremony, she showed interest in me and said that she was here to guide me and take me to some of the villages in the jungle. Was she more than my guide in the physical plane? Was she a messenger from the father who blows the winds of ayahuasca?

After the ceremonies with Don Solon in Iquitos, I had felt that this was enough ayahuasca for me. But, in spite of that, I had made the decision to follow Jorge and Don Solon back to Tarapoto to continue for more sessions. Three flights were canceled by "bad weather" that kept me from returning to Tarapoto for those sessions, which I feared I wouldn't be able to tolerate. Was the spirit wind of ayahuasca saving me? Had it given me enough?

It all fits together in some way that can only be seen from hindsight. The fact that my wandering in the jungle was the same as the wandering in the jungle of the film's young hero tells me that it was written. The fact that the film captured and brought into focus the very things I did in my own journey into the winds of ayahuasca tells me that on some level of reality, a mythic level, it was written. It's as if all the people I met were simply just there to guide me so that I could say my lines properly.

Why? What was my real reason for coming down here? I now knew. It was to discover the truth in the world of the shaman. It was to discover the other time, the mythic time of the shamanic universe. I found it and realized that it existed everywhere, although it hardly exists in the Western world which runs so strongly by chronological time. But it, nevertheless, exists. It exists in the mythic world of magic, of the unconscious. The reason I came was to reenchant myself, and once having done that, like the god Hermes, I was to bring the message back to my world.

I had been in the mythic world from the moment I entered Peru. Down here, not only did the oceans circulate in the opposite direction from the northern circulation, but my own system also began to circulate counter to the northern chronological stream of consciousness that is marked, milestoned, by each passing second of time.

It was a three-dimensional holographic movie starring me. I was scripted; my actions were determined; every event was planned for, but not in the ordinary world of time. It was in the parallel world of myth.

All of my quantum-theorizing said that nothing is written, nothing is determined. But that is only a half-truth. It is true in the world of logical time. It is not true in the world of mythic time. Suddenly the words of Dr. Cabieses appeared in mind. He had perhaps unwittingly told me that there are two worlds, two world views.

"You may feel that you created God, or you may feel that God created you; it doesn't matter. This is semantics."

It was more than semantics. It was the difference between the temporal and the mythic sense of time. In the temporal, man creates god. In the mythic, God creates man. Thus my sense of everything being written down, everything scripted, my destiny, the synchronicities of the events in the film and the events of my life, they also belonged to my mythic world. The other things that are important to me, such as time and money, and even explaining this to you, are part of the temporal, the so-called real world.

But which is real? In the temporal world we are ruled by the uncertainty principle of Werner Heisenberg. We cannot predict ahead of ourselves. We really don't know what anything is or really if anything is really there. There is no basis in solid reality for our experiences of solid reality. Yet we obviously sense that there is some objective world out there.

In the mythic world, a world that has escaped us since the industrial revolution, we have only glimpses of a deeper truth. We see that, as I saw, our lives do have meaning and they seem to be destined in a way that eludes us in the temporal world. Only at those rare times of great insight does this awareness come upon us.

Even Einstein had it half-right when he said that "God does not play dice with the universe." He was referring to the totally determined world of myth. He was wrong if he was talking about the temporal world. There he must write, "Man does play dice with the universe." That world of clock time is ruled by Heisenberg, not God.

God is the scriptwriter, but the script exists in the mythic world, the world we find meaning in. Men and women exist in both worlds, but they are able to move from one to the other and they are able to remain in one or the other. Stay in the mythic world while everyone else lives in the temporal world and you are found crazy and probably locked up. Or you become a great visionary, painter, shaman, or writer. You won't be elected president, however. Stay in the temporal world while everyone else lives in their mythic reality, and you are burned at the stake as a heretic.

But which world is the real world? I remember the words of Dr. Cabieses: "I'm thinking that the problem is that you are examining that . . . dilemma with your consciousness. But this is another dilemma. Actually it may be that you are inventing the whole thing. There may be no such a thing. So it is impossible to grasp."

But I didn't believe this. I had grasped something even if it wasn't anything material. I had seen beyond the temporal, searchlight activity of my own brain. It didn't matter to me whether these visions were already in my brain or not.

The announcement came over the loudspeaker. My plane was ready. I got up wearily and made my way on board. Suddenly I felt very old. My mind drifted back to my unconscious scab again. Growing older I sometimes fear the scab will continue to grow and protect me until no light will be seen. I will become "*tio*," an old man, as the spirits of Tarapoto called me, a man who dies. But when death comes and the scab is totally removed, perhaps I will be free to flow back to the Father and finally have an answer to the question, "Why do we have to remain here?"

I found my seat. I looked around the plane and noticed that one of the passengers, a Peruvian national that I had met on my flight out of Miami to Lima, was also aboard. I asked him about his stay in Lima. He originally told me that he was returning to Lima for good, so I was surprised to see him on this flight. I asked him what had happened. He said, "I don't know. As soon as I got home, I knew I had to return to the United States. Must be my fate, I guess. Who knows?"

I knew that I didn't, and for some perhaps bizarre reason I didn't care. I knew that I would be okay.

As the plane lifted off the ground, I looked back as the Lima coastline vanished from view. I looked around at my fellow passengers and I felt at ease. I knew that we were spiritual beings living in a material world. We had chosen to have this experience for only one reason: to remember who we really were, spiritual beings from a mythos who have chosen to experience time.

I had remembered who I really was and I finally knew why I had to remain here.

Bibliography

America Indigena: Instituto Indigenista Intramericano. Vol. XLVI. Mexico: Instituto Indigenista Interamericano, Insurgentes sur No. 1690, Colonia Florida Mexico, D.F., 1986.

Andrews, Lynn V. *Medicine Woman*. New York: Harper & Row, 1981.

"Angels, Aliens, and Archetypes." *Revision*. Volume 11, Numbers 3 and 4, Parts one and two, 1989.

Aspect, Alain; Dalibard, Jean; and Roger, Gerard. "Experimental Test of Bell's Inequalities Using Time-Varying Analyzers." *Physical Review Letters*. Vol. 49, No. 25, p. 1804, 20 December, 1982.

Bates, Brian. *The Way of Wyrd: Tales of an Anglo-Saxon Sorcerer*. New York: Berkley Books, 1988.

Bateson, Gregory. *Mind and Nature: A Necessary Unity*. New York: E.P. Dutton, 1979.

Becker, Robert O., and Selden, Gary. *The Body Electric*. New York: Morrow, 1985.

Bell, John Stewart. *Speakable and Unspeakable in Quantum Mechanics*. Cambridge, England: Cambridge University Press, 1987.

Bohm, David. *Wholeness and the Implicate Order*. Boston: Routledge & Kegan Paul, 1980.

Borges, Jorge. *Ficciones*. New York: Grove Press, 1962.

Boulle, Pierre. *Time Out of Mind*. New York: Signet Books, 1966.

Brown, Joseph Epes. *The Spiritual Legacy of the American Indian*. New York: Crossroad, 1988.

Buksbazen, John Daishin. *To Forget the Self*. Vol. III. Zen Writings Series. Los Angeles: Zen Center of Los Angeles, no date, p. 2.

Castañeda, Carlos. *The Power of Silence: Further Lessons of Don Juan*. New York: Simon & Schuster, 1987.

Cramer, John G. "Generalized Absorber Theory and the Einstein-Podolsky-Rosen Paradox." *Physical Review D*. Vol. 22, p. 362, 1980.

———. "Alternate Universes II." *Analog*. November, 1984.

———. "The Transactional Interpretation of Quantum Mechanics." *Reviews of Modern Physics*. Vol. 58, No. 3, July, 1986.

———. "The Quantum Handshake." *Analog*. November, 1986.

———. "An Overview of the Transactional Interpretation of Quantum Mechanics." *International Journal of Theoretical Physics*. Vol. 27, No. 2, pp. 227–236, 1988.

David-Neel, Alexandria. *Buddhism: Its Doctrines and Its Methods*. New York: Avon Books, 1979, pp. 37–41.

Davis, Wade. *The Serpent & the Rainbow*. New York: Simon & Schuster, 1985.

Devereux, Paul. *Earth Lights: Towards an Understanding of the UFO Enigma*. United Kingdom: Wellingborough, Northamptonshire: Turnstone, 1982.

———. *Earth Lights Revelation*. London: Blandford, 1989.

———. *Places of Power*. London: Blandford, 1990.

Devereux, Paul, and Pennick, Nigel. *Lines on the Landscape*. London: Robert Hale, 1989.

Devereux, Paul; Steele, John; and Kubrin, David. *Earthmind—Is the Earth Alive?* New York: Harper & Row, 1989.

Doherty, J. "Hot Feat: Firewalkers of the World." *Science Digest*. August, 1982, pp. 67–71. Also see the letter to the editor in response to this article, "Atheist Firewalkers." *Science Digest*. November, 1982, p. 11.

Doore, Gary, editor. *Shaman's Path: Healing, Personal Growth, and Empowerment*. Boston & London: Shambhala, 1988.

Easwaran, Eknath. *Dialogue with Death: The Spiritual Psychology of the Katha Upanishad*. Berkeley, California: The Blue Mountain Center of Mediation, 1981.

Eccles, J.C. "Do mental events cause neural events analogously to the probability fields of quantum mechanics?" *Proc. R. Soc.* Vol. B 227, pp. 411–428, 1986.

Einstein, Albert; Podolsky, Boris; and Rosen, Nathan. "Can The Quantum-Mechanical Description of Physical Reality Be Considered Complete?" *Physical Review*. Vol. 47, p. 777, 1935.

Eliade, Mircea. *Shamanism: Archaic Techniques of Ecstasy*. New Jersey: Bollingen Series LXXVI, Princeton University Press, 1964.

Ellis-Davidson, H.R. *The Gods and Myths of Northern Europe*. London, England: Penguin, 1977, pp. 66–69.

Fadiman, James, and Frager, Robert. *Personality and Personal Growth*. New York: Harper & Row, 1976, p. 358.

Gerber, Richard. *Vibrational Medicine: New Choices for Healing Ourselves*. Boulder, Colorado: Bear & Co., 1988.

Grof, Stanislav. *Beyond the Brain: Birth, Death, and Transcendence in Psychotherapy*. Albany, New York: State University of New York, 1985.

Halifax, Joan. *Shamanic Voices: A Survey of Visionary Narratives*. New York: E.P. Dutton, 1979.

Harner, Michael. *The Way of the Shaman: A Guide to Power and Healing*. New York: Harper & Row, 1980. Also New York: Bantam Books, 1982.

———. "What Is a Shaman," in Doore, Gary, editor. *Shaman's Path: Healing, Personal Growth, and Empowerment*. Boston & London: Shambhala, 1988, p. 7.

Heisenberg, Werner. *Physics and Beyond*. New York: Harper & Row, 1971.

Hellmuth, T.; Zajonc, Arthur C.; and Walther, H. "Realizations of Delayed Choice Experiments," in *New Techniques and Ideas in Quantum Measurement Theory*, ed. D.M. Greenberger, Vol. 480. Annals of the New York Academy of Sciences, December 30, 1986.

Highwater, Jamake. *The Sun, He Dies*. New York: Harper & Row, 1980.

———. *The Primal Mind: Visions and Reality in Indian America*. New York: Harper & Row, 1981.

Honderich, T. "The Time of a Conscious Sensory Experience and Mind-Brain Theories," *J. Theor. Biol.*, Vol. 110, pp. 115–119, 1984.

Hopkins, Bud. *Intruders*. New York: Random House, 1987.

Jaynes, Julian. *The Origin of Consciousness and the Breakdown of the Bicameral Mind*. Boston: Houghton Mifflin Co., 1976.

Kalweit, Holger. *Dreamtime & Inner Space: The World of the Shaman*. Boston: Shambhala, 1988.

Lamb, F. Bruce. *Wizard of the Upper Amazon: The Story of Manuel Cordova-Rios*. Berkeley, California: North Atlantic Books, 1971, 1974.

———. *Rio Tigre and Beyond: The Amazon Jungle Medicine of Manuel Cordova*. Berkeley, California, 1985.

Libet, B. "Subjective Antedating of a Sensory Experience and Mind-Brain Theories: Reply to Honderich (1984)." *J. Theor. Biol.*, Vol. 114, pp. 563–570, 1985.

Libet, B.; Wright, E.W.; Feinstein, B.; and Pearl, D.K. *Brain*, Vol. 102, p. 193, 1979.

Luna, Luis Eduardo. *Vegatalismo: Shamanism Among the Mestizo Population of the Peruvian Amazon*. Stockholm, Sweden: The University of Stockholm. 1983.

Mabit, Jacques Michel. "L'Hallucination par l'Ayahuasca Chez les Guérisseurs de la Haute-Amazonie Peruvienne (Tarapoto)." *Institut Français d'Etudes Andines*. (IFEA, Casilla 278, Lima 18.) France: Document de Travail, 1/1988.

Morris, W., editor. *The American Heritage Dictionary of the English Language*. Boston: American Heritage Publishing Co. and Houghton Mifflin Co., 1969.

Naranjo, Plutarco, "Ayahuasca in Archeological Ecuador." *America Indigena: Instituto Indigenista Intramericano*. Vol. XLVI. Mexico: Instituto Indigenista Interamericano, Insurgentes sur No. 1690, Colonia Florida Mexico, D.F., 1986.

Neihardt (Flaming Rainbow), John G. *Black Elk Speaks: Being the Life of a Holy Man of the Oglala Sioux*. Lincoln and London: University of Nebraska Press, First Bison Book Edition, 1988.

Pattee, Rowena. "Ecstasy and Sacrifice," in Doore, Gary, editor. *Shaman's Path: Healing, Personal Growth, and Empowerment*. Boston & London: Shambhala, 1988.

Puharich, Andrija. *Beyond Telepathy*. New York: Doubleday Anchor, 1973.

Quasha, George. "Aufzeichnung einer rede von Essie Parrish, gehalten am 14. März 1972." *Alcheringa*, No. 1, p. 27, 1975.

Rachowiecki, Rob. *Peru: A Survival Kit*. Berkeley, California: Lonely Planet Publications, 1987.

Rutherford, Ward. *Shamanism: The Foundations of Magic*. Wellingborough, Northamptonshire, England: The Aquarian Press, 1986.

The Sioux. A brochure distributed in South Dakota by the U.S. Department of the Interior, Indian Crafts Board, Sioux Indian Museum and Crafts Center, P.O. Box 1504, Rapid City, South Dakota.

Strieber, Whitley. *Communion: A True Story*. New York: Beech Tree Books, William Morrow, 1987.

———. *Transformation: The Breakthrough*. New York: Beech Tree Books, William Morrow, 1988.

Suarés, Carlo. *The Qabala Trilogy: The Cipher of Genesis; The Song of Songs; and The Sepher Yetsira*. Boston & London: Shambhala, 1985.

Tacitus. *Germania*, vol. 31.

Toben, Bob, and Wolf, Fred Alan. *Space-Time and Beyond*: The New Edition. New York: E.P. Dutton, 1982. Also New York: Bantam Books, 1983.

Vonnegut, Kurt. *Slaughterhouse Five*. New York: Dell Publishing Co., 1969.

Watson, Lyall. *Lightning Bird: One Man's Journey into Africa*. New York: Simon & Schuster, 1982.

Weil, Andrew. *Health and Healing. Understanding Conventional and Alternative Medicine*. Boston: Houghton Mifflin Co., 1983.

Weil, Pierre. "Vers une approche holistique de la nature de la réalité." *Médicines Nouvelles et Psychologies Transpersonnelles. L'Ouvert, Question de No. 64*, pp. 11–57, 1986. Albin Michel, publisher.

Wheeler, John A. "The Mystery and the Message of the Quantum." Presentation at the *Joint Annual Meeting of the American Physical Society and the American Association of Physics Teachers*. Jan., 1984.

Wolf, Fred Alan. *Taking the Quantum Leap: The New Physics for Nonscientists*. San Francisco: Harper & Row, 1981.

———. *Star*Wave: Mind, Consciousness, and Quantum Physics*. New York: Macmillan, 1984.

———. "The Quantum Physics of Consciousness: Towards a New Psychology." *Integrative Psychology*. Vol. 3, pp. 236–247, 1985.

———. *The Body Quantum: The New Physics of Body, Mind, and Health*. New York: Macmillan, 1986.

———. *Parallel Universes: The Search for Other Worlds*. New York: Simon & Schuster, 1989.

———. "On the Quantum Physical Theory of Subjective Antedating." *Journal of Theoretical Biology*. Vol. 136, pp. 13–19, 1989.

Zaleski, Carol. *Otherworld Journeys: Accounts of Near-Death Experiences in Medieval and Modern Times*. New York: Oxford Univ. Press, 1987.

Index

About the Author

FRED ALAN WOLF, a Ph.D. in theoretical physics, is the author of many books, including *Parallel Universes* and the American Book Award-winning *Taking the Quantum Leap*. He lives near Seattle, Washington.